The WHO Application of ICD-10 to deaths during pregnancy, childbirth and the puerperium: ICD-MM

WHO Library Cataloguing-in-Publication Data

The WHO application of ICD-10 to deaths during pregnancy, childbirth and puerperium: ICD MM.

1.Maternal mortality – classification. 2.Cause of death – classification. 3.Postpartum period. 4.Parturition. 5.Pregnancy complications – classification. 6.Pregnancy outcome. 7.Classification. 8.Manuals. I.World Health Organization.

ISBN 978 92 4 154845 8 (NLM classification: WQ 270)

© **World Health Organization 2012**

All rights reserved. Publications of the World Health Organization are available on the WHO web site (www.who.int) or can be purchased from WHO Press, World Health Organization, 20 Avenue Appia, 1211 Geneva 27, Switzerland (tel.: +41 22 791 3264; fax: +41 22 791 4857; e-mail: bookorders@who.int).

Requests for permission to reproduce or translate WHO publications – whether for sale or for noncommercial distribution – should be addressed to WHO Press through the WHO web site (http://www.who.int/about/licensing/copyright_form/en/index.html).

The designations employed and the presentation of the material in this publication do not imply the expression of any opinion whatsoever on the part of the World Health Organization concerning the legal status of any country, territory, city or area or of its authorities, or concerning the delimitation of its frontiers or boundaries. Dotted lines on maps represent approximate border lines for which there may not yet be full agreement.

The mention of specific companies or of certain manufacturers' products does not imply that they are endorsed or recommended by the World Health Organization in preference to others of a similar nature that are not mentioned. Errors and omissions excepted, the names of proprietary products are distinguished by initial capital letters.

All reasonable precautions have been taken by the World Health Organization to verify the information contained in this publication. However, the published material is being distributed without warranty of any kind, either expressed or implied. The responsibility for the interpretation and use of the material lies with the reader. In no event shall the World Health Organization be liable for damages arising from its use.

Printed in France

Acknowledgements

The WHO *Application of ICD-10 to deaths during pregnancy, childbirth and the puerperium* was developed by the WHO Working Group on Maternal Mortality and Morbidity Classification. The following individuals (listed in alphabetical order) participated in the activities of the WHO Working Group on Maternal Mortality and Morbidity Classification: Linda Bartlett, Jon Barrett, Alma Virginia Camacho, José Guilherme Cecatti, Veronique Filippi, Rogelio Gonzalez, Ahmet Metin Gülmezoglu, Anoma Jayathilaka, Affette McCaw-Binns, Robert C Pattinson, Mohamed Cherine Ramadan, Cleone Rooney, Lale Say, João Paulo Souza, Mary Ellen Stanton, Buyanjargal Yadamsuren and Nynke van den Broek, and Zelka Zupan. We thank numerous reviewers for critically reviewing the earlier drafts.

Robert Pattinson and Lale Say prepared the alpha and beta drafts of this work, based on the guidance provided by the working group. Lale Say, Robert Pattinson, Affette McCaw-Binns, João Paulo Souza and Cleo Rooney revised the beta draft, which was approved by the working group. The final version of the document was prepared by Doris Chou, Robert Pattinson, Cynthia Pileggi, Cleone Rooney and Lale Say.

We thank the Child Health Epidemiology Reference Group (CHERG), Robert Jakob, Patricia Wood, and the Mortality Reference Group of the ICD, and Maria Rodriguez for their technical review and comments of this work.

This work was funded by USAID, the UNDP/UNFPA/WHO/World Bank Special Programme of Research, Development and Research Training in Human Reproduction (HRP), and by a grant from the Bill and Melinda Gates Foundation to the US Fund for UNICEF for the work of the Child Health Epidemiology Reference Group.

Contents

Acknowledgements	iii
Abbreviations and acronyms	vi
Executive summary	vii
Introduction	1
Development of the WHO *Application of ICD-10 to deaths during pregnancy, childbirth, and the puerperium*	3
The WHO *Application of ICD-10 to deaths during pregnancy, childbirth and the puerperium*	7
Specific explanations and motivations	17
Implications for practice and research	21
Conclusion	21
References	22
Appendix 1: Reviewers of draft versions of the Classification of maternal mortality and morbidity	23
Annex A: List of codes and ICD-MM groups	24
Annex B1: Tabular List of ICD-10 codes that describe conditions which may be causes of death (underlying cause)	24
Group 1: Pregnancy with abortive outcome	25
Group 2: Hypertensive disorders in pregnancy, childbirth and the puerperium	28
Group 3: Obstetric Haemorrhage	29
Group 4: Pregnancy-related infection	32
Group 5: Other obstetric complications	34
Direct deaths without an Obstetric code in ICD-10	37
Group 6: Unanticipated complications of management	40
Group 7: Non-obstetric complications	42
Group 8: Unknown/undetermined	47
Group 9: Coincidental causes	47
Annex B2: Tabular List of Chapter 15 codes that describe conditions which are unlikely to cause death but may have contributed to the death (contributory condition)	48
Annex B3: Tabular List of Other codes of interest	65
Annex C: Suggestions of tools and examples to facilitate the implementation of the guide and its groupings	66

Abbreviations and acronyms

AFLP	acute fatty liver of pregnancy
AIDS	acquired immunodeficiency syndrome
APH	antepartum haemorrhage
CHERG	Child Health Epidemiology Reference Group
FIGO	Federation of Gynecology and Obstetrics
HELLP	haemolysis, elevated liver enzymes, low platelet count
HIV	human immunodeficiency virus
HRP	UNDP/UNFPA/WHO/World Bank Special Programme of Research Development and Research Training in Human Reproduction
ICD	International Statistical Classification of Diseases and Related Health Problems
MDG	Millennium Development Goal
NEC	not elsewhere classified
NOS	not otherwise specified
PPH	postpartum haemorrhage
PV	per vaginam
UNDP	United Nations Development Programme
UNFPA	United Nations Population Fund
UNICEF	United Nations Children's Fund
USAID	United States Agency for International Development
VR	vital registration
WHO	World Health Organization

Executive summary

Reducing maternal mortality by 75% is the Millennium Development Goal 5a. To reach this goal, countries need an accurate picture of the causes and levels of maternal deaths. However, efforts to document the progress in decreasing maternal mortality must make adjustments for inconsistencies in country-reported maternal mortality. Completeness of maternal death reporting and accuracy of statements of causes of death need to be improved and may compromise the output resulting from subsequent standardized coding and classification according to the rules of the International Statistical Classification of Diseases (ICD).

The WHO *Application of ICD-10 to deaths during pregnancy, childbirth, and the puerperium: ICD-Maternal Mortality* (ICD-MM) is based upon the 10th revision of the ICD (ICD-10) and its coding rules. It is intended to facilitate the consistent collection, analysis and interpretation of information on maternal deaths. Improved reporting will also facilitate the coding of conditions. This document is primarily intended to assist health-care providers, those who complete death certification by clarifying the application of the ICD-10 and standardizing the identification of direct and indirect maternal deaths. Its principles should be applicable for categorizing deaths data collected through civil registration, surveys, hospital information systems, verbal autopsies, confidential enquiries and other special studies.

The accompanying appendices and tables

- facilitate consistent reporting of the clinical conditions,
- identify conditions and codes which are unlikely causes of death but may have contributed to death,
- indicate which causes of death are counted as direct or indirect maternal deaths.

Ultimately, standardization of the cause of death attribution will improve:

- interpretation of data on maternal mortality,
- analysis on the causes of maternal death,
- allocation of resources and programmes intended to address maternal mortality.

Applying ICD-MM will decrease errors in coding and improve cause of maternal death attribution. This will enhance usability and comparability of maternal mortality statistics generated from ICD data. It is recommended that countries adopt the ICD-MM, and statistical offices and academicians collect data according to the ICD-MM.

The guide should always be used in conjunction with the three volumes of ICD-10. The suggested code should be verified and possible additional information should be coded using the full ICD-10, Volumes 1 and 3; rules for selection of underlying cause of death and certification of death apply in the way they are described in ICD-10 Volume 2.

The WHO Application of ICD-10 to deaths during pregnancy, childbirth and the puerperium: ICD MM

Introduction

Reducing maternal mortality is one of the key targets within the Millennium Development Goals (MDGs). To reach this target, countries need an accurate picture of the levels and causes of maternal deaths (*1*). A majority of countries use the International statistical classification of diseases and related health problems, Tenth revision (ICD-10) as the standard tool to guide their collection, coding, tabulation and reporting of mortality statistics based on civil registration(*2*).

In the ICD-10, deaths with a causal and/or temporal relationship to pregnancy are characterized and defined as maternal deaths due to direct or indirect causes, deaths during pregnancy, childbirth and puerperium, or late maternal deaths (see Box 3). Despite guidance within the ICD and definitions that describe discrete entities, in practice, the identification, reporting and consequent classification of maternal death are inconsistent (*3*). There remains apparent confusion between symptoms, signs and diseases, and which conditions should be reported and accordingly tabulated as cause of death. The reporting also impacts on the ability of coding to either indirect maternal or incidentally maternal deaths. An analysis of cause of maternal death data found variation in the way deaths are reported in different countries (*3*).

A range of conditions that are frequently reported have different public health impact in view of progress in measures to improve pregnancy outcomes and reducing the maternal mortality such as obstructed labour, anaemia, or HIV. Specific rationale and explanations and motivation for their revised handling are given later in this document (page 21).

An immediate consequence of the inconsistency in death attribution, reporting, and resulting coding, is misclassification and underreporting of maternal deaths extracted from vital registration (VR), which in turn may bias understanding of the magnitude and causes of maternal death (*4–13*). The implication of this bias on programmatic work and public health policies then becomes readily apparent. Recognizing the particular difficulty in identifying maternal deaths, the 43rd World Health Assembly in 1990 approved the addition of a "checkbox" to ICD death certificates to indicate whether a woman was pregnant, or had recently terminated/delivered a pregnancy at the time (*14*). This was incorporated into ICD-10 Volume 2 (*2*) and implemented in more than 30 countries (*15*).

In response to the ongoing need for a better understanding of the underlying causes of death, WHO initiated an activity aiming to develop, test and promote standardization of reporting and new ways of tabulating maternal causes of death, in line with ICD-10. The Application of ICD-10 to deaths in childbirth, pregnancy, and the puerperium is based upon the 10th revision of the ICD (ICD-10) and follows all rules for mortality coding as described in Volume 2 of the ICD. The application clarifies relevance of existing ICD-10 codes and related conditions and provides guidance to meaningful grouping of ICD categories to enable consistent application of ICD coding and rules to improve data collection and analysis.

This document presents:

- a brief summary of the development of this guide;
- a grouping system for identification of maternal deaths using existing ICD-10 codes, which countries can immediately implement.

This document is intended to be used by those charged with death certification. It is intended to guide their ability to document the pertinent information by clarifying which conditions should be considered underlying causes of death; thus, improving accurate death attribution. As a result, the information available to coders, programme managers, statistical offices, and academicians/researchers will be improved.

The WHO Application of ICD-10 to deaths during pregnancy, childbirth and the puerperium: ICD MM

Development of the WHO *Application of ICD-10 to deaths during pregnancy, childbirth, and the puerperium*

The guide and groupings described here, based on ICD-10, were developed through a consultative process. WHO established a technical working group of obstetricians, midwives, epidemiologists and public health professionals from developing and developed countries to prepare this standard guide for capturing information relating to deaths during pregnancy, childbirth and the puerperium. The group adopted three principles for its work. First, the new guide and groupings should be practical and understood by its users (clinicians, coders, epidemiologists, programme managers, and researchers). Second, in line with ICD rules, detailed underlying cause categories should be mutually exclusive and should identify all of the conditions that are epidemiologically and/or clinically important. The clinically related conditions are aggregated in new groups that facilitate epidemiological analysis and health service planning and evaluation. Third, the way of grouping that results form this work does contribute to and will be compatible with the 11th revision of the ICD.

An alpha-draft of this guide, groupings, and the recommendations for classification was peer reviewed by more than 40 individuals, professional societies (e.g. the International Federation of Gynecology and Obstetrics (FIGO), the Royal College of Obstetricians and Gynaecologists, the American College of Obstetricians and Gynecologists, and the Canadian College of Obstetricians and Gynaecologists) and relevant international agencies. Following this feedback, a second version was tested on nine databases of maternal deaths: national registration and surveillance databases from Colombia, Jamaica, Mongolia and South Africa; other health facility based databases from Kenya, Malawi and Zimbabwe; and verbal autopsy data from Afghanistan and Nigeria. This was performed following the steps described in Box 1.

Based on the experiences accumulated during the test with databases, and the input received from experts, a revised, beta-draft recommendation for groupings was prepared. This was reviewed

---

Box 1

Steps taken for testing the guide

1. Identification and description of the denominator population.

2. Verification and description of data-collection procedures and methods (for the original set).

3. Assignation of causes of deaths by using the new groupings with respect to underlying causes and contributory conditions.

4. Determination of the proportion (%) of deaths that could not be classified with the new system, and determination of the reasons for nonclassification (e.g. category does not exist, category now within contributory factors and the real cause cannot be identified).

5. Comparison of differences on the distribution of causes as compared to previous attribution.

6. Assessment of the difficulty/ease of using the proposed system.

7. Identification of specific issues that would require prospective study.

by a wide range of stakeholders for further inputs and revision, and finalized by WHO. Interactions with the ICD Secretariat and the ICD-11 revision team were held in order to ensure consistency and compatibility between the proposed guide and the ICD.

This document is based upon ICD-10 codes and coding principles. However, in the course of this work, needs for additional and different detail not reflected in ICD-10 were identified, resulting in proposals for new codes to be included in the ICD-11. Also the proposed groupings of categories shade a new light on needs of public health in maternal mortality and needs for future changes to ICD. As the revision towards ICD-11 is ongoing, the reader is referred to the web page available at http://www.who.int/classifications/icd/revision/en/index.html for further details on the process of revisions and suggested changes to the ICD. Once ICD-11 is released, any new codes pertinent to cause of death attribution in pregnancy, childbirth or the puerperium, will be added to future updates of this guide.

The WHO Application of ICD-10 to deaths during pregnancy, childbirth and the puerperium: ICD MM

The WHO *Application of ICD-10 to deaths during pregnancy, childbirth and the puerperium*

Understanding death certification, ICD terms and relationship to maternal deaths

Cause of death: documentation and analysis

Certification of cause of death

Cause of death is first determined by the certifier who reports the morbid conditions and events leading to the woman's death on a medical certificate of cause of death. It is essential that at this stage all relevant information is reported in a complete fashion. ICD-10 lays out the format of the medical certificate of cause of death, which is designed to help the certifier record the whole sequence of events leading to death in Part 1, in steps starting from the immediate cause on line 1a and going back to each earlier step on subsequent lines (top to bottom) until they get to the earliest event, usually the underlying cause. Part 1 should always include clear information about whether mutual aggravation between a disease and pregnancy lead to death (indirect maternal deaths).

Based on the ICD recommendations, countries produce their own forms for use in civil registration and provide accompanying instructions to certifiers/doctors on how to complete them. Based upon a resolution approved by the 43rd World Health Assembly (WHA 43.24), ICD-10 recommends that countries should consider inclusion on death certificates questions about current pregnancy and pregnancy within one year preceding death (ICD-10 VOL 2 para 5.8.1). This has been shown to reduce underreporting of maternal deaths (*16*). It reminds the certifier to consider whether the death was due to a complication of pregnancy. Figure 1 gives an example of the medical certification of cause of death (MCCD).

Figure 1. Example of the medical certification of cause of death (MCCD).

Cause of death *the disease or condition thought to be the underlying cause should appear in the lowest completed line of part I*		Approximate interval between onset and death
Part I Disease or condition leading directly to death a)		
Antecedent causes: Due to or as a consequence of b)		
Due to or as a consequence of c)		
Due to or as a consequence of d)		
Part II Other significant conditions Contributing to death but not related to the disease or condition causing it		
The woman was: ☐ pregnant at the time of death ☐ not pregnant at the time of death (but pregnant within 42 days) ☐ pregnant within the past year		

Countries may add tick boxes to the form of medical certificate of cause of death (MCCD) to indicate pregnancy.

Coding cause of death

A trained coder then codes the conditions mentioned on the death certificate and after applying the ICD-10 rules for coding and selection assigns a single ICD-10 code for the single underlying cause of death. The pregnancy tick box informs the coder to consider whether the death might be coded to a maternal death. For indirect maternal deaths, it is essential that in Part 1 of the certificate there is a clear statement about mutual aggravation between the pregnancy and the disease leading to death.

Analysing cause of death

Statisticians or analysts then aggregate these ICD codes into epidemiologically and clinically meaningful groups and publish mortality statistics. This statistical information is used by multiple stakeholders, whose objectives may differ, but all users rely heavily on the *quality*, *accuracy* and *consistency* of the data.

Box 2

ICD-10 terminology

Underlying cause of death is defined as the disease or condition that initiated the morbid chain of events leading to death or the circumstances of the accident or violence that produced a fatal injury. The single identified cause of death should be as specific as possible.

If the death certificate has been completed correctly, the underlying cause of death should normally be the single condition which the certifier has written on the lowest used line of Part 1. The mortality selection and modification rules in Volume 2 of ICD-10 have been developed to enable coders to select the most useful information on cause of death for public health purposes as the single underlying cause, even when the certificate is not completed correctly or where it is important to consider/combine information from other parts of the certificate.

In multiple cause coding, all of the conditions on the death certificate are assigned ICD codes and retained for statistical analysis. These include other contributory conditions in the sequence in Part 1 and conditions in Part 2. *Contributory* cause is used here to include conditions that may exist prior to development of the underlying cause of death or develop during the chain of events leading to death and which, by its nature, contributed to the death. In this document however, contributory conditions also refers to conditions that might be reported in Part 1 of the certificate.[1]

Interested readers are also referred to detailed training on ICD which can be found online apps.who.int/classifications/apps/icd/icd10training/ or downloaded for offline use: apps.who.int/classifications/apps/icd/ClassificationDownload/DLArea/OfflineTrainingPackage.zip

This WHO guide and its revised groupings of maternal deaths were developed in relation to mortality data derived from civil registration with medical certification of cause of death. However, it can be used in other settings, e.g. where the cause of death is determined by verbal autopsy, survey or confidential enquiry.

Using existing ICD-10 codes, this document identifies those conditions that may be a potential cause of death and are of high public health and distinguishes them from those that are unlikely to cause death but may have contributed to or been part of the course of events leading to death.

Irrespective of setting, the guide and its groupings have been devised to capture at least the most important basic information on cause of death, while allowing for refinement with more specific details. At the most basic level, 'Deaths during pregnancy, childbirth or the puerperium' may be enumerated even in countries or areas where no information on cause of death is available. Mortality rates can then be compared with those based on data aggregated across all causes in areas where cause is available.

---

[1] In this document contributory causes refers to conditions that may be reported in Part 1 of the death certificate. These are also referred to as "intervening causes" within ICD terminology,

> **Box 3**
>
> **Definition of deaths in pregnancy, childbirth and the puerperium: ICD-10**
>
> *Death occurring during pregnancy, childbirth and the puerperium* is the death of a woman while pregnant or within 42 days of termination of pregnancy, irrespective of the cause of death (obstetric and non-obstetric).
>
> **Maternal death**
> A maternal death is the death of a woman while pregnant or within 42 days of termination of pregnancy, irrespective of the duration and the site of the pregnancy, from any cause related to or aggravated by the pregnancy or its management, but not from accidental or incidental causes.
>
> Maternal deaths are subdivided into two groups:
> - *direct obstetric deaths:* direct obstetric deaths are those resulting from obstetric complications of the pregnancy state (pregnancy, labour and the puerperium), from interventions, omissions, incorrect treatment, or from a chain of events resulting from any of the above.
>
> - *indirect obstetric deaths:* indirect obstetric deaths are those resulting from previous existing disease or disease that developed during pregnancy and which was not due to direct obstetric causes, but which was aggravated by physiologic effects of pregnancy.
>
> **Late maternal death**
> A late maternal death is the death of a woman from direct or indirect causes more than 42 days but less than one year after termination of pregnancy.

Application of ICD-10 to deaths during pregnancy, childbirth and the puerperium

This document standardizes identifying relevant causes of death and ensures their accurate reporting. In such way conditions can be coded in a more detailed way and the quality of information related to maternal death (see Box 3) will improve. With training on the rationale of death certification and how the derivative data are used, certifiers will be better able to complete death certificates with meaningful data. ICD coding rules are not affected by the re-grouping of ICD codes, and in fact the standardization of maternal underlying causes of death codes ensures that ICD coding rules are followed. In countries that collect VR data based on medical certification, trained coders code the cause of death to the highest level of detail per ICD-10 coding convention.

In settings where the cause of death is identified by verbal autopsy or similar data gathering from reporters who have not been trained in clinical diagnosis or certification of cause of death, it may only be possible to classify causes of death to relatively broad groups. It has often been necessary for clinicians to reformulate the histories from lay reporters into sequences in the ICD death certificate format to identify the underlying cause even at this broad group level.

Annex A provides an electronic link to an excel sheet indicating the group for each existing ICD-10 code in Chapter XV. Additionally, tools to assist in the implementation of this guide and its groupings and to synergize with maternal death review and audit processes are also in development.

Analysis of underlying causes of death

In order to foster a common framework for international comparisons, categories of underlying causes of death were aggregated in nine groups of causes of death during pregnancy, childbirth and the puerperium. These groups are clinically and epidemiologically relevant, mutually exclusive and totally inclusive and descriptive of all causes of maternal and pregnancy-related deaths. Furthermore, they simplify the characterization of maternal deaths, whether due to direct and indirect causes.[2]

Table 1 presents the nine groups of causes during pregnancy, childbirth and the puerperium, with examples of corresponding conditions to be included in each group. Clinically, conditions that may result in mortality may also cause morbidity and these specified as conditions that should be identified as underlying cause of maternal deaths. A complete listing of conditions that may be underlying causes of either death or maternal morbidity is detailed in Annexes B1, B2, and B3.

In some settings, the underlying cause of death may only be identified at the broad level of the group, whereas in other areas, the cause of death may be attributed with more detail, at category or subcategory level. In practice, consistent allocation of deaths to broad groups may be more difficult than actual consistent coding to detailed ICD codes and subsequent aggregation into larger groups. In either case, it is essential to have a good understanding of the meaning of terms used in that setting to describe cause of death and accurate and consistent indexing of all such terms to the correct category at whatever level of detail is in use. Note that any local modifications of the nine groups into categories and subcategories will not affect the overall standardization of attribution of cause of death or its classification and definition as a "maternal death", or "death during pregnancy, childbirth and the puerperium".

---

[2] An important rationale for creating these groupings is to clarify and standardize the reporting of conditions considered to have high public health impact. The coding of the identified conditions, correctly filled in on death certificates, follows coding procedures described in ICD-10 Volume 2. From the perspective of analysis, these are labelled *single underlying cause of death* as consistent with the ICD.

Table 1

Groups of underlying causes of death during pregnancy, childbirth and the puerperium in mutually exclusive, totally inclusive groups [3]

Type	Group name/number	EXAMPLES of potential causes of death
Maternal death: direct	1. Pregnancies with abortive outcome	Abortion, miscarriage, ectopic pregnancy and other conditions leading to maternal death and a pregnancy with abortive outcome
Maternal death: direct	2. Hypertensive disorders in pregnancy, childbirth, and the puerperium	Oedema, proteinuria and hypertensive disorders in pregnancy, childbirth and the puerperium
Maternal death: direct	3. Obstetric haemorrhage	Obstetric diseases or conditions directly associated with haemorrhage
Maternal death: direct	4. Pregnancy-related infection	Pregnancy-related, infection-based diseases or conditions
Maternal death: direct	5. Other obstetric complications	All other direct obstetric conditions not included in groups to 1–4
Maternal death: direct	6. Unanticipated complications of management	Severe adverse effects and other unanticipated complications of medical and surgical care during pregnancy, childbirth or the puerperium
Maternal death: indirect	7. Non-obstetric complications	Non-obstetric conditions • Cardiac disease (including pre-existing hypertension) • Endocrine conditions • Gastrointestinal tract conditions • Central nervous system conditions • Respiratory conditions • Genitourinary conditions • Autoimmune disorders • Skeletal diseases • Psychiatric disorders • Neoplasms • Infections that are not a direct result of pregnancy
Maternal death: unspecified	8. Unknown/undetermined	Maternal death during pregnancy, childbirth and the puerperium where the underlying cause is unknown or was not determined
Death during pregnancy, childbirth and the puerperium	9. Coincidental causes	Death during pregnancy, childbirth and the puerperium due to external causes

[3] See Annex A and B1 for complete enumeration and details

Conditions unlikely to cause death but may have contributed to the events leading to death
(Contributory conditions)

The section describes conditions that may have contributed to or may be associated with, but should not to be reported as sole condition on the death certificate or selected as the underlying cause of death. *Contributing causes may predispose women to death, as either a pre-existing condition or a risk factor.* For example, in a woman with twin gestation, whose delivery is complicated by uterine atony and postpartum bleeding, hypovolaemic shock, disseminated intravascular coagulopathy and renal failure. In this case, using multiple cause coding, the contributory conditions include twin gestation (ICD code O30.0), shock, DIC, and renal failure whereas the underlying cause of death is postpartum haemorrhage resulting from uterine atony (ICD-10 code O72.1). If only single cause coding is used, only the underlying cause of death, postpartum haemorrhage (uterine atony), O72.1 would be recorded.

Annex B2 presents a separate tabular list of the conditions unlikely to cause death, these 'contributory' codes that may be used in multiple cause of death coding to describe maternal morbidities associated with pregnancy, childbirth or the puerperium. It is possible that more than one contributory condition may exist, and in this circumstance, multiple coding for these conditions is recommended. These codes are not to be selected as underlying cause of death, because they do not capture the most useful information needed for health service and public health interventions to prevent further deaths.

In the example above, the other diagnoses of hypovolaemic shock, disseminated intravascular coagulopathy and renal failure are complications, and these are indicated in Part 1 of the death certificate. It is necessary to document the complications that resulted in the death, as this might help in developing treatment protocols to prevent such complications in the future. Further, a pattern can be detected that may help in the management of similar women in the future. Complications encompass significant morbidities such as organ system dysfunction, and the codes for these conditions are found in the morbidity list.

Annexes B1 and B2 presents clinically and epidemiologically relevant conditions to be considered as possible morbidities.

Applicability: The following examples are intended illustrate the format of death certificate completion, documenting the sequence of events from the underlying cause to the immediate cause of death and the feasibility of applying the groupings in practice.

EXAMPLE 1

A woman who had anaemia during pregnancy and after delivery had a postpartum haemorrhage due to uterine atony, and died as a result of hypovolaemic shock.

Medical certificate of cause of death

Cause of death *the disease or condition thought to be the underlying cause should appear in the lowest completed line of Part I*		Approximate interval between onset and death
1. Disease or condition leading directly to death	(a) hypovolaemic shock — *A contributory cause indicated in Part 1. This is assigned a code when multiple cause coding is undertaken*	10 minutes
Antecedent causes: Due to or as a consequence of	(b) postpartum haemorrhage	30 minutes
Due to or as a consequence of	(c) uterine atony — *The underlying cause. This is the last condition noted in Part 1 and is a condition found in Annex B1*	45 minutes
Due to or as a consequence of	(d)	
2. Other significant conditions Contributing to death but not related to the disease or condition causing it	Anaemia	pre-existing
The woman was: ☒ pregnant at the time of death ☐ not pregnant at the time of death (but pregnant within 42 days) ☐ pregnant within the past year		

If deceased was a woman, was she pregnant when she died or within 42 days before she died ? Yes

(Part I shaded for purposes of the example)

EXAMPLE 2

A woman infected with HIV who has a spontaneous abortion that becomes infected, and dies due to septic shock and renal failure.

Medical certificate of cause of death

Cause of death *the disease or condition thought to be the underlying cause should appear in the lowest completed line of Part I*		Approximate interval between onset and death
1. Disease or condition leading directly to death	(a) renal failure — *A contributory condition, indicated in Part 1*	2 hours
Antecedent causes: Due to or as a consequence of	(b) septic shock	24 hours
Due to or as a consequence of	(c) septic miscarriage — *The underlying cause. This is the last condition noted in Part 1 and is a condition found in Annex B*	36 hours
Due to or as a consequence of	(d)	
2. Other significant conditions Contributing to death but not related to the disease or condition causing it	HIV — *A contributory condition, indicated in Part IIB*	pre-existing
The woman was: ☒ pregnant at the time of death ☐ not pregnant at the time of death (but pregnant within 42 days) ☐ pregnant within the past year		

If deceased was a woman, was she pregnant when she died or within 42 days before she died ? Yes

(Part I shaded for purposes of the example)

Verbal autopsy

In some settings, maternal deaths are ascertained by verbal autopsy. Once the relevant details are extracted from the verbal autopsy, the maternal death guide and its groupings may also be used to standardize the information regarding cause of death, see Example 3 (*17*).

EXAMPLE 3

This was the woman's third pregnancy and she had not had any complications during the first two deliveries. She did not have any tetanus toxoid vaccination or antenatal consultation by a doctor or nurse in any of her pregnancies because of her religious beliefs. She ate normally and her health was good, although she sometimes suffered from headaches at which time she liked to lie down on her bed. After six months of pregnancy she became unable to see at night but no consultation with a doctor was arranged for this problem. She did not develop any bodily swelling.

When she was nine months pregnant, it was one day before her death, she went into labour at about 7 o'clock in the evening and she called her mother (who was a Dai) to the house. After she had finished her "esha" prayer at 9 o'clock in the evening, her labour pain increased a little. Her mother examined her and felt that the baby's head was not yet close to the birth passage. At 11 o'clock at night her mother examined her again and found slight vaginal bleeding. She examined her a total of three times.

Around midnight, her labour pain increased again and after another hour her waters broke. After fifteen minutes of watery discharge, she had a normal delivery at 1 o'clock at night. After five minutes, her placenta was also normally delivered. During the delivery she had normal blood and water discharge.

Shortly after the delivery, she said that she felt dizzy and wanted to lie down. Her mother-in-law and sister-in-law (husband's brother's wife) were washing her baby. Suddenly she said she felt sick, she developed a headache and wanted to sit down. As soon as her sister–in-law helped her to sit on the bed she developed excessive vaginal bleeding. She then stood up on a jute mat, which became soaked with blood. After that, she was made to lie down but she still had excessive bleeding, which continued for another hour.

Her husband tried to fetch a doctor but he said he would not come until the morning. After an hour of excessive PV [per vaginam] bleeding, the woman's whole body had become cold and pale. The bleeding then began to slow down and she was given hot compresses. However, some time after the bleeding had reduced, she began to tremble and started to clench her teeth. After 30 minutes in that condition she became exhausted and remained on the bed with her eyes closed. At 5 o'clock in the early morning she had three hiccups and died.

Medical certificate of cause of death

Cause of death *the disease or condition thought to be the underlying cause should appear in the lowest completed line of Part I*		Approximate interval between onset and death
1. Disease or condition leading directly to death	(a) postpartum haemorrhage	3 hours
Antecedent causes: Due to or as a consequence of	(b)	
Due to or as a consequence of	(c)	
Due to or as a consequence of	(d)	
2. Other significant conditions Contributing to death but not related to the disease or condition causing it	Lack of access to medical care to prevent or treat haemorrhage following normal vaginal delivery	
The woman was: ☒ pregnant at the time of death ☐ not pregnant at the time of death (but pregnant within 42 days) ☐ pregnant within the past year		

Callouts:
- The underlying cause. This is the last condition noted in Part 1 and is a condition found in Annex B
- A contributory condition, indicated in Part II. No code is assigned because only single cause coding of deaths

If deceased was a woman, was she pregnant when she died or within 42 days before she died ? Yes

(Part I shaded for purposes of the example)

The WHO Application of ICD-10 to deaths during pregnancy, childbirth and the puerperium: ICD MM

Specific explanations and motivations

Prolonged/obstructed labour

The ICD-10 aims to capture the initiating step most relevant to public health in the sequence leading to death, because preventing this condition would prevent not just the death, but all of the illness, complications and disability that preceded it. Obstructed labour may be the start of a sequence leading to death, or may itself be due to some preceding condition such as contracted maternal pelvis or transverse fetal lie. In these cases, death might be prevented by access to operative delivery. However, there is evidence that many deaths may be mis-attributed to obstructed labour, leading to over-estimation of the proportion that could be prevented through operative delivery and underestimating the need for other services. In areas where deliveries are not attended by trained professionals and maternal mortality is high, very little information may be available about the sequence of events that lead to death, or about the progress of labour. The only information from lay reporters may be that the woman appeared to be in labour, or in pain, for a considerable time before death, and/ or that she died undelivered. These deaths may then be attributed to obstructed labour, without any good evidence that the condition really existed.

The WHO working group decided that it would be preferable only to accept the diagnosis if further evidence, for example the fatal complication of obstructed labour (e.g. ruptured uterus, uterine atony/haemorrhage or sepsis) was specified. In other words, certifiers should report more detail on the death certificate, than just obstructed labour. It should also be noted that this decision reflects the principles used to develop the groupings and the recommendations of the ICD, e.g., that the identified underlying cause must be mutually exclusive. The use of obstructed labour as an underlying cause alone is not sufficient, as exemplified by the case of uterine rupture associated with obstructed labour.

In this clinical scenario, there are two conditions but only the ruptured uterus can be considered the single underlying cause whereas obstructed labour may have multiple clinical outcomes as it not only contributes to the ruptured uterus but also other conditions such as puerperal sepsis.

At present, some countries report obstructed labour as a contributing condition while other countries report obstructed labour and ruptured uterus as causes of death. It is important to standardize this in order to permit informed analysis of comparable data on causes of death. Programmatically, the objective is to prevent obstructed labour, and, when not possible or once obstructed labour is diagnosed, the need is to identify the access to emergency obstetric care and the allocation of services (e.g. access to safe blood transfusion, antibiotics and postpartum care in the event of fistulas).

In practice, in settings where mortality is covered by vital registration, individual countries will be able to disaggregate national data by both underlying cause and contributory causes, where multiple cause reporting and analysis is feasible, ensuring that no loss of information occurs. In setting where information on maternal mortality is collected by other mechanisms such as maternal death audit, maternal death review or verbal autopsy, certifiers of death are informed by this guide that obstructed labour alone is insufficient as a cause of death, it is envisaged that they will be prompted to supply more information about the circumstances of death. As a result, programmes should be able to identify additional health interventions, such as access to safe blood transfusion and antibiotics, needed to prevent these deaths.

This additional level of detail is feasible and will increase the robustness of information available to programme managers and policy makers who are in the position to influence the quality and availability of care to avoid preventable maternal deaths.

This recommendation is predicated on the need for training of certifiers of death (health-care providers) to understand that a diagnosis of obstructed labour alone is usually insufficient . Where multiple cause coding is undertaken, the specification of this detail will be easily incorporated. However, in the circumstance that single cause of death coding is performed, if details surrounding the death of a woman who was diagnosed with obstructed labour are provided, these deaths will be coded as seen in Example 4. If no other information is provided (see Example 5), then coders would be obligated to use the codes for obstructed labour as the underlying cause of death. In both circumstances, the death is considered a direct maternal death, counting each case of obstructed labour, but only Example 4 provides details on potential gaps in intrapartum care provision.

It is important to note that this description of reconciling data regarding obstructed labour is particular to ICD-10. With future revisions of ICD, it is anticipated that coding for obstructed labour and its associated conditions (e.g., haemorrhage, sepsis) will be simplified with the proposal of new linked codes that identify both concepts in one code and streamline single cause of death coding.

EXAMPLE 4

This was the woman's third pregnancy and she had not had any complications during the first two deliveries. She did not have any tetanus toxoid vaccination or antenatal consultation by a doctor or nurse in any of her pregnancies because of her religious beliefs. She ate normally and her health was good, although she sometimes suffered from headaches at which time she liked to lie down on her bed. After six months of pregnancy she became unable to see at night but no consultation with a doctor was arranged for this problem. She did not develop any bodily swelling.

A woman with a baby in breech position who experiences obstructed labour and dies of puerperal sepsis

- Underlying cause: Group 4, pregnancy-related infection
- Category: puerperal sepsis
- Contributing condition: obstructed labour due to fetal malpresentation

EXAMPLE 5

A woman who dies very soon after arriving at a health facility. She died undelivered, but health personnel at the facility are able to feel fetal parts on vaginal examination. The person accompanying her to the health facility is only able to indicate that she had "pains" for more than a day and a half.

- Underlying cause: Group 5, other obstetric complications
- Category: obstructed labour NOS (not otherwise specified)
- Contributing condition: no details

This change will indicate:
- the number of deaths that follow the development of obstructed labour,
- the number of women who die of conditions amenable to treatment such as blood transfusions or antibiotics, which will inform programmes on areas of need in the antenatal and intrapartum period.

HIV and AIDS

There is a tendency in many parts of the world to attribute all deaths in people known to have HIV or AIDS to AIDS. However, such patients may die "from AIDS", or "with HIV". Temporal to pregnancy, it is useful to distinguish those deaths of HIV-infected women that should be considered maternal deaths.

In terms of dying "with HIV" or "from AIDS", women may die from obstetric causes, e.g. incomplete abortion, complicated by haemorrhage or tetanus, or an ectopic pregnancy. These deaths are considered direct maternal deaths. In these cases, their HIV infection or AIDS might have coexisted at time of death but it is not the underlying cause of death.

In contrast, "AIDS related indirect maternal deaths" are deaths of HIV-infected women who die because of the aggravating effect of pregnancy on HIV. This interaction between pregnancy and HIV is the underlying cause of death. These are coded as O98.7 and categorized in Group 7 (non-obstetric complications). Proper reporting of the mutual influence of HIV or AIDS and pregnancy in Part 1 of the certificate will guide the coders.

On the other hand, a woman with HIV may die of one of the fatal complications of HIV or AIDS while pregnant, though this is probably a rare event since such severe illness makes pregnancy unlikely. An example may be when an HIV-positive woman who is in early pregnancy dies due to HIV wasting syndrome. Here the pregnancy is incidental to her underlying cause of death, which is HIV wasting syndrome. In these rare cases, HIV or AIDS is selected as the underlying cause of death and the appropriate code in block B20-B24 of ICD-10 selected. These are termed "HIV-related deaths to women during pregnancy, delivery or puerperium" and are *not* considered maternal deaths.

Classifying each and every case in terms of HIV status will give a clearer picture of the role of HIV and AIDS in maternal deaths. The convention of using O98.7 to describe indirect maternal deaths and appropriate B codes to describe deaths of women when HIV or AIDS is the underlying cause and where pregnancy is incidental will reduce confusion and standardize statistical tabulation.

Anaemia

With the exception of pre-existing disease such as sickle cell disease, or thalassaemia, anaemia may be secondary to infections, malnutrition, bleeding, etc. Anaemia rarely causes death on its own. In this guide and its groupings, anaemia is a factor contributing to maternal death. Even where anaemia complicates postpartum haemorrhage, it is still almost always the haemorrhage that caused the death.

Tetanus

OB tetanus (ICD 10 code A34) is a rare cause of maternal death. For the purposes of classification, in the absence of detailed information regarding the clinical course of infection, it is considered an DIRECT maternal cause of death within the group "pregnancy related infection". Where there is evidence that tetanus exposure and infection is the result of an obstetric event, eg abortion or puerperal sepsis, the death is classified to the respective DIRECT cause of death.

Malnutrition

This is not a disease entity causing death, but may have contributed to the death.

Female genital mutilation

This is common in some areas of the world and may contribute the death of a woman due to the scarring causing prolonged labour and predisposing the women to uterine atony, puerperal sepsis or severe lower genital tract trauma due to tearing of the scar tissue.

Previous caesarean section

This may have contributed to the death by promoting placenta accreta, uterine rupture or placenta praevia.

Obesity, depression and domestic violence

Obesity is becoming an increasing problem, and by facilitating collection of these data on maternal deaths the impact of obesity on maternal deaths might be better understood. The same is true for depression and domestic violence as contributory conditions but it may be more difficult to collect the information on every case.

Suicide

The ICD-10 and ICD-MM recommend collecting all pertinent information describing the events leading to death. Within ICD-10 coding convention, maternal deaths due to suicide and coded appropriately to the Chapter XX within vital registration data alone would not be considered within international maternal mortality estimation per current methodology.[4] However, when maternal deaths due to suicide are included within surveillance reporting, these would be included in the maternal mortality estimation studies dataset.

Antenatal and postpartum suicide are grouped in this guide under direct causes of death under the "Other" category. This is recommended even if it may not be possible to definitively establish the diagnosis of puerperal psychosis and/or postpartum depression. ICD-10 categorizes suicides in the code range X60-X84 in the Chapter XX. Hence, maternal deaths due to suicide are identified when information about the pregnancy was indicated on the death certificate, either in Part 1, 2, or by the "tick box".

Specific to late postpartum suicide occurring between 42 days and one year postpartum, these may receive an additional code O96.0 (late maternal death from direct obstetric cause) or if greater than one year postpartum. If the event occurs greater than one year postpartum and an established diagnosis of puerperal psychosis and/or postpartum depression exists, these may receive an additional code as death from sequelae of direct obstetric cause (O97.0).[5]

---

[4] The Maternal Mortality estimation Interagency Group publishes updates on global maternal mortality. The methodology describing how maternal deaths are identified within vital registration data can be found in the full report, *Trends in maternal mortality: 1990 to 2010 WHO, UNICEF, UNFPA and The World Bank estimates* (http://www.who.int/reproductivehealth/publications/monitoring/9789241503631/en/index.html) and ICD-10 Volume 2.

[5] For single condition coding and tabulation of the cause of death, the external cause describing the suicide should be used as the primary code. For multiple coding, the additional information on pregnancy related depression and the differentiation between late maternal and sequelae will be useful in the analysis of maternal causes of death.

Figure 2. All deaths (death during pregnancy, childbirth or puerperium)

ALL DEATHS
(death during pregnancy, childbirth or puerperium)

MATERNAL DEATH			OTHER DEATHS
Direct maternal death: • abortive outcome • hypertensive disorders • obstetric haemorrhage • pregnancy related infection • other obstetric complications • unanticipated complications	**In-direct maternal death:** • non-obstetric complications	**Unknown Undetermined**	**Coincidental**

Implications for practice and research

The guide and its groupings are expected to render a better assessment of conditions leading to death during pregnancy, childbirth and the puerperium. Applying this guide and its groupings should help to identify the real clinical causes and health-system shortfalls that countries need to address in order to reduce complications and fatal outcomes of pregnancy. Annex C provides additional suggestions of tools to facilitate implementation. The use of this guide and its groupings are recommended as part of the efforts to estimate and address the burden of maternal mortality around the world.

Conclusion

The WHO Application of ICD-10 to deaths during pregnancy, childbirth and the puerperium builds upon the ICD-10 to create a useful framework for programme officers, health-care workers certifying deaths, and statistical offices and researchers. It has the potential to improve the quality of data derived from all sources of information on the cause of maternal death. This will improve comparability of data and inform the development of programmes to decrease maternal mortality.

Since the guide and its groupings build upon the ICD, end users will be familiar with the clinical concepts organized within the groupings of this guide. The advantage of this guide lies in that simplicity. Future research on the application of the guide and its groupings is necessary.

Achieving MDG5 will require an understanding of not only the magnitude but also the contribution of causes of death. Currently, about one third of WHO Member States/territories are able to provide high-quality VR data. Even so, it is recognized that misclassification of deaths that are temporal to pregnancy occurs in these data. The use of the guide, in addition to a pregnancy check box on death certificates, is intended to improve the accurate capture of data and its attribution.

For Member States and territories with some facility to capture VR events, the guide is poised to improve their quality of VR data on attribution of cause. Where data are collected by means of special surveys, the *Application of ICD-10 to Maternal Mortality: ICDMM* will improve comparability of data.

References

1. *Trends in maternal mortality: 1990 to 2008. Estimates developed by WHO, UNICEF, UNFPA, and The World Bank*. Geneva, World Health Organization, 2010.

2. *World Health Organization: International statistical classification of diseases and related health problems, 10th revision*. Geneva, World Health Organization, 1992.

3. Daniels J, et al. The WHO analysis of causes of maternal death, in preparation, 2011.

4. Karimian-Teherani D et al. Under-reporting of direct and indirect obstetrical deaths in Austria, 1980–98. *Acta Obstetrica et Gynecologica Scandinavica*, 2002; 81:323–327.

5. Bouvier-Colle MH et al. Reasons for the underreporting of maternal mortality in France, as indicated by a survey of all deaths among women of childbearing age. *International Journal of Epidemiology*, 1991, 20(3):717–721.

6. Deneux-Tharaux C et al. Underreporting of pregnancy-related mortality in the United States and Europe. *Obstetrics and Gynecology*, 2005, 106(4):684–692. Erratum in: *Obstetrics and Gynecology*, 2006, 107(1):209.

7. Gissler M et al. Pregnancy-related deaths in four regions of Europe and the United States in 1999–2000: characterisation of unreported deaths. *European Journal of Obstetrics Gynecology and Reproductive Biology*, 2007, 133(2):179–185.

8. Schuitemaker N et al. Underreporting of maternal mortality in The Netherlands. *Obstetrics and Gynecology*, 1997, 90(1):78–82.

9. Kao S et al. Underreporting and misclassification of maternal mortality in Taiwan. *Acta Obstetrica et Gynecologica Scandinavica*, 1997, 76(7):629–636.

10. Department of Health. *Report on confidential enquiries into maternal deaths in the United Kingdom 1988–1990*. London, HMSO, 1994.

11. Horon IL. Underreporting of maternal deaths on death certificates and the magnitude of the problem of maternal mortality. *American Journal of Public Health*, 2005, 95(3):478–482.

12. Salanave B et al. Classification differences and maternal mortality: a European study. MOMS Group. Mothers' Mortality and Severe morbidity. *International Journal of Epidemiology*, 1999, 28(1):64–69.

13. Atrash HK, Alexander S, Berg CJ. Maternal mortality in developed countries: not just a concern of the past. *Obstetrics and Gynecology*, 1995, 86(4 Pt 2):700–705.

14. Resolution WHA 43.24. Report of the International Conference for the Tenth Revision of the International Classification of Diseases. In Forty-third World Health Assembly, Geneva. (Fourteenth plenary meeting, 17 May 1990, Committee B, third report.

15. Reported information on mortality statistics. Geneva, World Health Organization, 2005. (http://www.who.int/healthinfo/mort2005survey/en/index.html, accessed 8 March 2012)

16. Horon IL, Cheng D. Effectiveness of pregnancy check boxes on death certificates in identifying pregnancy-associated mortality. *Public Health Report*, 2011, 126(2):195–200.

17. *Why mothers die in Matlab*. Dhaka, ICDDRB, Centre for Health and Population Research, 2005. (http://centre.icddrb.org/images/Why_mothers_die_in_Matlab2.pdf, accessed 8 March 2012).

Appendix 1: Reviewers of draft versions of the Classification of maternal mortality and morbidity

Dorothy Shaw, FIGO

Margaret Wash, FIGO

Barbara de Zalduondo, UNAIDS

Francisco Songane, Partnership for Maternal, Newborn and Child Health

Gwyneth Lewis, Department of Health, United Kingdom

Luc de Bernis, UNFPA

Vincent Fauveau, UNFPA

Wendy Graham, Initiative for Maternal Mortality Programme Assessment

Zoe Matthews, Department for International Development, United Kingdom

Julia Hossein, Initiative for Maternal Mortality Programme Assessment

Kathy Herschderfer, International Confederation of Midwives

Country reviewers

Guillermo Carroli, Argentina

Jose Guilherme Cecatti, Brazil

Anibal Faundes, Brazil

Zhao-Gengli, China

Edgar Kestler, Guatemala

Sunita Mittal, India

Manorama-Balkisan Purwar, India

Horace Fletcher, Jamaica

Cherry Than-Than-Tin, Myanmar

Prasanna-Gunasekera, Nepal

Saramma T. Mathai, Nepal

Mario Festin, Philippines

Thilina Palihawadana, Sri Lanka

Prof. H.R. Seneviratne, Sri Lanka

Pisake Lumbiganon, Thailand

Sompop Limpongsanurak, Thailand

Tippawan Tippawan-Liabsuetrakul, Thailand

Jose Villar, United Kingdom

Alain Prual, United States

Tran Son Thach, Viet Nam

Regional Advisers

WHO Regional Office for Africa
Seipati Mothebesoane Anoh
Djamila Cabral

WHO Regional Office for the Americas
Ricardo Fescina
Bremen de Mucio

WHO Regional Office for South-East Asia
Ardi Kaptiningsih

WHO Regional Office for Europe
Gunta Lazdana
Alberta Bacci

WHO Regional Office for the Eastern Mediterranean
Ramez Mahaini,
Hossam Mahmoud

WHO Regional Office for the Western Pacific
Narimah Awin

Annex A: List of codes and ICD-MM groups

A complete listing of ICD-10 codes and corresponding ICD-MM groups can be found online www.who.int/reproductivehealth/publications/monitoring/9789241548458/en/

This list should always be used in conjunction with the three volumes of ICD-10. The suggested code should be verified and possible additional information should be coded using the full ICD-10, looking up the terms in Volume 3 and verifying the code with Volume 1; rules for certification and selection of the underlying cause of death apply in the way they are described in ICD-10 Volume 2.

Annex B1: Tabular List of ICD-10 codes that describe conditions which may causes of death (underlying cause)

Codes in this section may be used in mortality or morbidity coding (unless the code is specifically specified as a mortality code)

Codes are grouped into the nine groups of obstetric causes of death, rather than the order of the tabular list in Volume 1 of the ICD-10, or its special tabulation lists and may not contain all codes within a block.

For the purposes of this guide, only conditions and associated codes in this section should be selected as underlying causes of deaths.

The annex should always be used in conjunction with the three volumes of ICD-10. The suggested code should be verified and possible additional information should be coded using the full ICD-10, Volumes 3 and 1; rules for certification of death apply in the way they are described in ICD-10 Volume 2. Further modifications as published in the 11th revision of the ICD may result in changes.

Group 1: Pregnancy with abortive outcome

Excl.: continuing pregnancy in multiple gestation after abortion of one fetus or more (O31.1)
The following fourth-character subdivisions are for use with categories O03-O06:
Note: Incomplete abortion includes retained products of conception following abortion.

.0 Incomplete, complicated by genital tract and pelvic infection
With conditions in O08.0

.1 Incomplete, complicated by delayed or excessive haemorrhage
With conditions in O08.1

.2 Incomplete, complicated by embolism
With conditions in O08.2

.3 Incomplete, with other and unspecified complications
With conditions in O08.3-O08.9

.4 Incomplete, without complication

.5 Complete or unspecified, complicated by genital tract and pelvic infection
With conditions in O08.0

.6 Complete or unspecified, complicated by delayed or excessive haemorrhage
With conditions in O08.1

.7 Complete or unspecified, complicated by embolism
With conditions in O08.2

.8 Complete or unspecified, with other and unspecified complications
With conditions in O08.3-O08.9

.9 Complete or unspecified, without complication

O00   Ectopic pregnancy

Incl.: ruptured ectopic pregnancy

Use additional code from category O08.-, if desired, to identify any associated complication.

O00.0   **Abdominal pregnancy**

Excl.: delivery of viable fetus in abdominal pregnancy (O83.3)

maternal care for viable fetus in abdominal pregnancy (O36.7)

O00.1   **Tubal pregnancy**

Fallopian pregnancy

Rupture of (fallopian) tube due to pregnancy

Tubal abortion

O00.2   **Ovarian pregnancy**

O00.8   **Other ectopic pregnancy**

Pregnancy:

- cervical
- cornual
- intraligamentous
- mural

O00.9   **Ectopic pregnancy, unspecified**

O01 Hydatidiform mole

Use additional code from category O08.-, if desired, to identify any associated complication.

Excl.: malignant hydatidiform mole (D39.2)

O01.0 Classical hydatidiform mole

Complete hydatidiform mole

O01.1 Incomplete and partial hydatidiform mole

O01.9 Hydatidiform mole, unspecified

Trophoblastic disease NOS

Vesicular mole NOS

O02 Other abnormal products of conception

Use additional code from category O08.-, if desired, to identify any associated complication.

Excl.: papyraceous fetus (O31.0)

O02.0 Blighted ovum and nonhydatidiform mole

Mole:

- carneous
- fleshy
- intrauterine NOS

Pathological ovum

O02.1 Missed abortion

Early fetal death with retention of dead fetus

Excl.: missed abortion with:

- blighted ovum (O02.0)
- mole:
- hydatidiform (O01.-)
- nonhydatidiform (O02.0)

O02.8 Other specified abnormal products of conception

Excl.: those with:

- blighted ovum (O02.0)
- mole:
- hydatidiform (O01.-)
- nonhydatidiform (O02.0)

O02.9 Abnormal product of conception, unspecified

O03 Spontaneous abortion

[See before O03 for subdivisions]

Incl.: miscarriage

O04 Medical abortion

[See before O03 for subdivisions]

Incl.: termination of pregnancy:
- legal
- therapeutic

therapeutic abortion

O05 Other abortion

[See before O03 for subdivisions]

O06 Unspecified abortion

[See before O03 for subdivisions]

Incl.: induced abortion NOS

O07 Failed attempted abortion

Incl.: failure of attempted induction of abortion

Excl.: incomplete abortion (O03-O06)

O07.0 Failed medical abortion, complicated by genital tract and pelvic infection

With conditions in O08.0

O07.1 Failed medical abortion, complicated by delayed or excessive haemorrhage

With conditions in O08.1

O07.2 Failed medical abortion, complicated by embolism

With conditions in O08.2

O07.3 Failed medical abortion, with other and unspecified complications

With conditions in O08.3-O08.9

O07.4 Failed medical abortion, without complication

Failed medical abortion NOS

O07.5 Other and unspecified failed attempted abortion, complicated by genital tract and pelvic infection

With conditions in O08.0

O07.6 Other and unspecified failed attempted abortion, complicated by delayed or excessive haemorrhage

With conditions in O08.1

O07.7 Other and unspecified failed attempted abortion, complicated by embolism

With conditions in O08.2

O07.8 Other and unspecified failed attempted abortion, with other and unspecified complications

With conditions in O08.3-O08.9

O07.9 Other and unspecified failed attempted abortion, without complication

Failed attempted abortion NOS

Group 2: Hypertensive disorders in pregnancy, childbirth and the puerperium

(note that O10, pre-existing hypertension is in Group 7)

O11 **Pre-existing hypertensive disorder with superimposed proteinuria**
Incl.: Conditions in O10.- complicated by increased proteinuria
Superimposed pre-eclampsia

O12 **Gestational [pregnancy-induced] oedema and proteinuria without hypertension**

O12.0 Gestational oedema

O12.1 Gestational proteinuria

O12.2 Gestational oedema with proteinuria

O13 **Gestational [pregnancy-induced] hypertension without significant proteinuria**
Incl.: Gestational hypertension NOS
Mild pre-eclampsia

O14 **Gestational [pregnancy-induced] hypertension with significant proteinuria**
Excl.: superimposed pre-eclampsia (O11)

O14.0 Moderate pre-eclampsia

O14.1 Severe pre-eclampsia

O14.2 HELLP syndrome
Combination of hemolysis, elevated liver enzymes and low platelet count

O14.9 Pre-eclampsia, unspecified

O15 **Eclampsia**
Incl.: convulsions following conditions in O10-O14 and O16
eclampsia with pregnancy-induced or pre-existing hypertension

O15.0 **Eclampsia in pregnancy**

O15.1 **Eclampsia in labour**

O15.2 **Eclampsia in the puerperium**

O15.9 **Eclampsia, unspecified as to time period**
Eclampsia NOS

O16 **Unspecified maternal hypertension**

Group 3: Obstetric Haemorrhage

O20   **Haemorrhage in early pregnancy**
Excl.: pregnancy with abortive outcome (O00-O08)

O20.0   **Threatened abortion**
Haemorrhage specified as due to threatened abortion

O20.8   **Other haemorrhage in early pregnancy**

O20.9   **Haemorrhage in early pregnancy, unspecified**

O43   **Placental disorders**
Excl.: maternal care for poor fetal growth due to placental insufficiency (O36.5)
placenta praevia (O44.-)
premature separation of placenta [abruptio placentae] (O45.-)

O43.2   **Morbidly adherent placenta**

O44   **Placenta praevia**

O44.1   **Placenta praevia with haemorrhage**
Low implantation of placenta, NOS or with haemorrhage
Placenta praevia:
- marginal
- partial
- total

NOS or with haemorrhage
Excl.: labour and delivery complicated by haemorrhage from vasa praevia (O69.4)

O45   **Premature separation of placenta [abruptio placentae]**

O45.0   **Premature separation of placenta with coagulation defect**
Abruptio placentae with (excessive) haemorrhage associated with:
- afibrinogenaemia
- disseminated intravascular coagulation
- hyperfibrinolysis
- hypofibrinogenaemia

O45.8   **Other premature separation of placenta**

O45.9   **Premature separation of placenta, unspecified**
Abruptio placentae NOS

O46   **Antepartum haemorrhage, not elsewhere classified**
Excl.: haemorrhage in early pregnancy (O20.-)
intrapartum haemorrhage NEC (O67.-)
placenta praevia (O44.-)
premature separation of placenta [abruptio placentae] (O45.-)

	O46.0	**Antepartum haemorrhage with coagulation defect**
		Antepartum haemorrhage (excessive) associated with:
		• afibrinogenaemia
		• disseminated intravascular coagulation
		• hyperfibrinolysis
		• hypofibrinogenaemia
	O46.8	**Other antepartum haemorrhage**
	O46.9	**Antepartum haemorrhage, unspecified**

O67 Labour and delivery complicated by intrapartum haemorrhage, not elsewhere classified

Excl.: antepartum haemorrhage NEC (O46.-)

placenta praevia (O44.-)

postpartum haemorrhage (O72.-)

premature separation of placenta [abruptio placentae] (O45.-)

O67.0	**Intrapartum haemorrhage with coagulation defect**
	Intrapartum haemorrhage (excessive) associated with:
	• afibrinogenaemia
	• disseminated intravascular coagulation
	• hyperfibrinolysis
	• hypofibrinogenaemia
O67.8	**Other intrapartum haemorrhage**
	Excessive intrapartum haemorrhage
O67.9	**Intrapartum haemorrhage, unspecified**
	Rupture of uterus not stated

O71 Other obstetric trauma

Incl.: damage from instruments

O71.0	**Rupture of uterus before onset of labour**
O71.1	**Rupture of uterus during labour**
	as occurring before onset of labour
O71.3	**Obstetric laceration of cervix**
	Annular detachment of cervix
O71.4	**Obstetric high vaginal laceration alone**
	Laceration of vaginal wall without mention of perineal laceration
	Excl.: with perineal laceration (O70.-)
O71.7	**Obstetric haematoma of pelvis**
	Obstetric haematoma of:
	• perineum
	• vagina
	• vulva

O72 Postpartum haemorrhage

Incl.: haemorrhage after delivery of fetus or infant

O72.0 **Third-stage haemorrhage**

Haemorrhage associated with retained, trapped or adherent placenta

Retained placenta NOS

Use additional code, if desired, to identify any morbidly adherent placenta (O43-O45)

O72.1 **Other immediate postpartum haemorrhage**

Haemorrhage following delivery of placenta

Postpartum haemorrhage (atonic) NOS

O72.2 **Delayed and secondary postpartum haemorrhage**

Haemorrhage associated with retained portions of placenta or membranes

Retained products of conception NOS, following delivery

O72.3 **Postpartum coagulation defects**

Postpartum:

- afibrinogenaemia
- fibrinolysis

Group 4: Pregnancy-related infection

O23   Infections of genitourinary tract in pregnancy

O23.0   Infections of kidney in pregnancy

O23.1   Infections of bladder in pregnancy

O23.2   Infections of urethra in pregnancy

O23.3   Infections of other parts of urinary tract in pregnancy

O23.4   Unspecified infection of urinary tract in pregnancy

O23.5   Infections of the genital tract in pregnancy

O23.9   Other and unspecified genitourinary tract infection in pregnancy

Genitourinary tract infection in pregnancy NOS

O41.1   Infection of amniotic sac and membranes

Amnionitis

Chorioamnionitis

Membranitis

Placentitis

O75.3   Other infection during labour

Sepsis during labour

O85   Puerperal sepsis

Incl.: Puerperal:

- endometritis
- fever
- peritonitis
- sepsis

Use additional code (B95-B98), if desired, to identify infectious agent.

Excl.: obstetric pyaemic and septic embolism (O88.3)

sepsis during labour (O75.3)

O86   Other puerperal infections

Use additional code (B95-B98), if desired, to identify infectious agent.

Excl.: infection during labour (O75.3)

O86.0   Infection of obstetric surgical wound

Infected:

- caesarean section wound following delivery
- perineal repair following delivery

O86.1   Other infection of genital tract following delivery

Cervicitis

Vaginitis following delivery

O86.2 **Urinary tract infection following delivery**
Conditions in N10-N12, N15.-, N30.-, N34.-, N39.0 following delivery

O86.3 **Other genitourinary tract infections following delivery**
Puerperal genitourinary tract infection NOS

O86.4 **Pyrexia of unknown origin following delivery**
Puerperal:
- infection NOS
- pyrexia NOS

Excl.: puerperal fever (O85)

pyrexia during labour (O75.2)

O86.8 **Other specified puerperal infections**

O91 Infections of breast associated with childbirth

Incl.: the listed conditions during pregnancy, the puerperium or lactation

O91.0 **Infection of nipple associated with childbirth**
Abscess of nipple:
- gestational
- puerperal

O91.1 **Abscess of breast associated with childbirth**
Mammary abscess

Purulent mastitis

Subareolar abscess

gestational or puerperal

O91.2 **Nonpurulent mastitis associated with childbirth**
Lymphangitis of breast

Mastitis:
- NOS
- interstitial
- parenchymatous

gestational or puerperal

Group 5: Other obstetric complications

O21.1	**Hyperemesis gravidarum with metabolic disturbance**

Hyperemesis gravidarum, starting before the end of the 22nd week of gestation, with metabolic disturbance such as:

- carbohydrate depletion
- dehydration
- electrolyte imbalance

O21.2	**Late vomiting of pregnancy**

Excessive vomiting starting after 22 completed weeks of gestation

O22   **Venous complications in pregnancy**

Excl.: obstetric pulmonary embolism (O88.-)

the listed conditions as complications of:

- abortion or ectopic or molar pregnancy (O00-O07 , O08.7)
- childbirth and the puerperium (O87.-)

O22.3	**Deep phlebothrombosis in pregnancy**

Deep-vein thrombosis, antepartum

O22.5	**Cerebral venous thrombosis in pregnancy**

Cerebrovenous sinus thrombosis in pregnancy

O22.8	**Other venous complications in pregnancy**
O22.9	**Venous complication in pregnancy, unspecified**

Gestational:

- phlebitis NOS
- phlebopathy NOS
- thrombosis NOS

O24	**Diabetes mellitus in pregnancy**

Incl.: in childbirth and the puerperium

O24.4	**Diabetes mellitus arising in pregnancy**

Gestational diabetes mellitus NOS

O26.6	**Liver disorders in pregnancy, childbirth and the puerperium**

Cholestasis (intrahepatic) in pregnancy

Obstetric cholestasis

Excl.: hepatorenal syndrome following labour and delivery (O90.4)

O26.9	**Pregnancy-related condition, unspecified**

O71   **Other obstetric trauma**

Incl.: damage from instruments

O71.2	**Postpartum inversion of uterus**

O71.5	**Other obstetric injury to pelvic organs**
	Obstetric injury to:
	• bladder
	• urethra
O71.6	**Obstetric damage to pelvic joints and ligaments**
	Avulsion of inner symphyseal cartilage
	Damage to coccyx
	Traumatic separation of symphysis (pubis)
	Obstetric
O71.8	**Other specified obstetric trauma**
O71.9	**Obstetric trauma, unspecified**

O73 Retained placenta and membranes, without haemorrhage

O73.0	**Retained placenta without haemorrhage**
	Use additional code, if desired, to identify any morbidly adherent placenta (O43–O45)
O73.1	**Retained portions of placenta and membranes, without haemorrhage**
	Retained products of conception following delivery, without haemorrhage
O75.4	**Other complications of obstetric surgery and procedures**
	Cardiac:
	• arrest
	• failure
	Cerebral anoxia
	following caesarean or other obstetric surgery or procedures, including delivery
	NOS
	Excl.: complications of anaesthesia during labour and delivery (O74.-)
	obstetric (surgical) wound:
	• disruption (O90.0–O90.1)
	• haematoma (O90.2)
	• infection (O86.0)
O75.8	**Other specified complications of labour and delivery**
O75.9	**Complication of labour and delivery, unspecified**

O87 Venous complications in the puerperium

Incl.: in labour, delivery and the puerperium

Excl.: obstetric embolism (O88.-)

venous complications in pregnancy (O22.-)

O87.1	**Deep phlebothrombosis in the puerperium**
	Deep-vein thrombosis, postpartum
	Pelvic thrombophlebitis, postpartum
O87.3	**Cerebral venous thrombosis in the puerperium**
	Cerebrovenous sinus thrombosis in the puerperium

O87.9   Venous complication in the puerperium, unspecified

Puerperal:

- phlebitis NOS
- phlebopathy NOS
- thrombosis NOS

O88   Obstetric embolism

Incl.: pulmonary emboli in pregnancy, childbirth or the puerperium

Excl.: embolism complicating abortion or ectopic or molar pregnancy (O00-O07 , O08.2)

O88.0   Obstetric air embolism

O88.1   Amniotic fluid embolism

Anaphylactoid syndrome of pregnancy

O88.2   Obstetric blood-clot embolism

Obstetric (pulmonary) embolism NOS

Puerperal (pulmonary) embolism NOS

O88.3   Obstetric pyaemic and septic embolism

O88.8   Other obstetric embolism

O90   Complications of the puerperium, not elsewhere classified

O90.0   Disruption of caesarean section wound

O90.1   Disruption of perineal obstetric wound

Disruption of wound of:

- episiotomy
- perineal laceration

Secondary perineal tear

O90.2   Haematoma of obstetric wound

O90.3   Cardiomyopathy in the puerperium

Conditions in I42.-

O90.4   Postpartum acute renal failure

Hepatorenal syndrome following labour and delivery

O90.5   Postpartum thyroiditis

O90.8   Other complications of the puerperium, not elsewhere classified

Placental polyp

O90.9   Complication of the puerperium, unspecified

Direct deaths without an Obstetric code in ICD-10

Note: It is recognized that establishing a link between puerperal psychosis or depression may not be possible however, when suicide occurs temporal to pregnancy, childbirth, and the puerperium, these deaths will be considered as direct maternal deaths.

At present time, these underlying causes of death do not have an "O" code in ICD-10, It is advised that certifiers indicate on the death certificate the pregnancy status in order to minimize underreporting of suicide in pregnancy.

(X60-X84)	**Intentional self-harm**
	Incl.: purposely self-inflicted poisoning or injury
	suicide (attempted)
X60	**Intentional self-poisoning by and exposure to nonopioid analgesics,**
	antipyretics and antirheumatics
	Incl.: 4-aminophenol derivatives
	nonsteroidal anti-inflammatory drugs [NSAID]
	pyrazolone derivatives
	salicylates

691

International Classification of Diseases - ICD-11 2010

X61	**Intentional self-poisoning by and exposure to antiepileptic, sedativehypnotic,**
	antiparkinsonism and psychotropic drugs, not elsewhere classified
	Incl.: antidepressants
	barbiturates
	hydantoin derivatives
	iminostilbenes
	methaqualone compounds
	neuroleptics
	psychostimulants
	succinimides and oxazolidinediones
	tranquillizers
X62	**Intentional self-poisoning by and exposure to narcotics and psychodysleptics [hallucinogens], not elsewhere classified**
	Incl.: cannabis (derivatives)
	cocaine
	codeine
	heroin
	lysergide [LSD]
	mescaline
	methadone
	morphine
	opium (alkaloids)

X63   **Intentional self-poisoning by and exposure to other drugs acting on the autonomic nervous system**

Incl.: parasympatholytics [anticholinergics and antimuscarinics] and spasmolytics

parasympathomimetics [cholinergics]

sympatholytics [antiadrenergics]

sympathomimetics [adrenergics]

X64   **Intentional self-poisoning by and exposure to other and unspecified drugs, medicaments and biological substances**

Incl.: agents primarily acting on smooth and skeletal muscles and the respiratory system

anaesthetics (general)(local)

drugs affecting the:

• cardiovascular system

• gastrointestinal system

hormones and synthetic substitutes

systemic and haematological agents

systemic antibiotics and other anti-infectives

therapeutic gases

topical preparations

vaccines

water-balance agents and drugs affecting mineral and uric acid metabolism

X65   **Intentional self-poisoning by and exposure to alcohol**

Incl.: alcohol:

• NOS

• butyl [1-butanol]

• ethyl [ethanol]

• isopropyl [2-propanol]

• methyl [methanol]

• propyl [1-propanol]

fusel oil

692

Chapter XX

X66   **Intentional self-poisoning by and exposure to organic solvents and halogenated hydrocarbons and their vapours**

Incl.: benzene and homologues

carbon tetrachloride [tetrachloromethane]

chlorofluorocarbons

petroleum (derivatives)

X67 Intentional self-poisoning by and exposure to other gases and vapours

Incl.: carbon monoxide

lacrimogenic gas [tear gas]

motor (vehicle) exhaust gas

nitrogen oxides

sulfur dioxide

utility gas

Excl.: metal fumes and vapours (X69)

X68	**Intentional self-poisoning by and exposure to pesticides**
	Incl.: fumigants
	fungicides
	herbicides
	insecticides
	rodenticides
	wood preservatives
	Excl.: plant foods and fertilizers (X69)
X69	**Intentional self-poisoning by and exposure to other and unspecified chemicals and noxious substances**
	Incl.: corrosive aromatics, acids and caustic alkalis
	glues and adhesives
	metals including fumes and vapours
	paints and dyes
	plant foods and fertilizers
	poisonous foodstuffs and poisonous plants
	soaps and detergents
X70	**Intentional self-harm by hanging, strangulation and suffocation**
X71	**Intentional self-harm by drowning and submersion**
X72	**Intentional self-harm by handgun discharge**
X73	**Intentional self-harm by rifle, shotgun and larger firearm discharge**
X74	**Intentional self-harm by other and unspecified firearm discharge**
X75	**Intentional self-harm by explosive material**
X76	**Intentional self-harm by smoke, fire and flames**
X77	**Intentional self-harm by steam, hot vapours and hot objects**
X78	**Intentional self-harm by sharp object**
X79	**Intentional self-harm by blunt object**
	693
	International Classification of Diseases - ICD-11 2010
X80	**Intentional self-harm by jumping from a high place**
	Incl.: intentional fall from one level to another
X81	**Intentional self-harm by jumping or lying before moving object**
X82	**Intentional self-harm by crashing of motor vehicle**
	Incl.: intentional collision with:
	• motor vehicle
	• train
	• tram (streetcar)
	Excl.: crashing of aircraft (X83)
X83	**Intentional self-harm by other specified means**
	Incl.: intentional self-harm by:
	• caustic substances, except poisoning
	• crashing of aircraft
	• electrocution
X84	**Intentional self-harm by unspecified means**

Group 6 : Unanticipated complications of management
Category: Anaesthesia

O29 **Complications of anaesthesia during pregnancy**

Incl.: maternal complications arising from the administration of a general or local anaesthetic, analgesic or other sedation during pregnancy

Excl.: complications of anaesthesia during:
- abortion or ectopic or molar pregnancy (O00-O08)
- labour and delivery (O74.-)
- puerperium (O89.-)

O29.0 Pulmonary complications of anaesthesia during pregnancy

Aspiration pneumonitis

Inhalation of stomach contents or secretions NOS

Mendelson's syndrome

Pressure collapse of lung

due to anaesthesia during pregnancy

O29.1 Cardiac complications of anaesthesia during pregnancy

Cardiac:
- arrest
- failure

due to anaesthesia during pregnancy

O29.2 Central nervous system complications of anaesthesia during pregnancy

Cerebral anoxia due to anaesthesia during pregnancy

O29.3 Toxic reaction to local anaesthesia during pregnancy

O29.5 Other complications of spinal and epidural anaesthesia during pregnancy

O29.6 Failed or difficult intubation during pregnancy

O29.8 Other complications of anaesthesia during pregnancy

O29.9 Complication of anaesthesia during pregnancy, unspecified

O74 **Complications of anaesthesia during labour and delivery**

Incl.: maternal complications arising from the administration of a general or local anaesthetic, analgesic or other sedation during labour and delivery

O74.0 Aspiration pneumonitis due to anaesthesia during labour and delivery

Inhalation of stomach contents or secretions NOS

Mendelson's syndrome due to anaesthesia during labour and delivery

O74.1 Other pulmonary complications of anaesthesia during labour and delivery

Pressure collapse of lung due to anaesthesia during labour and delivery

O74.2   **Cardiac complications of anaesthesia during labour and delivery**

Cardiac:

• arrest

• failure

due to anaesthesia during labour and delivery

O74.3   **Central nervous system complications of anaesthesia during labour and delivery**

Cerebral anoxia due to anaesthesia during labour and delivery

O74.4   **Toxic reaction to local anaesthesia during labour and delivery**

O74.6   **Other complications of spinal and epidural anaesthesia during labour and delivery**

O74.7   **Failed or difficult intubation during labour and delivery**

O74.8   **Other complications of anaesthesia during labour and delivery**

O74.9   **Complication of anaesthesia during labour and delivery, unspecified**

O89   Complications of anaesthesia during the puerperium

Incl.: maternal complications arising from the administration of a general or local anaesthetic, analgesic or other sedation during the puerperium

O89.0   **Pulmonary complications of anaesthesia during the puerperium**

Aspiration pneumonitis

Inhalation of stomach contents or secretions NOS

Mendelson's syndrome

Pressure collapse of lung

due to anaesthesia during the puerperium

O89.1   **Cardiac complications of anaesthesia during the puerperium**

Cardiac:

• arrest

• failure

due to anaesthesia during the puerperium

O89.2   **Central nervous system complications of anaesthesia during the puerperium**

Cerebral anoxia due to anaesthesia during the puerperium

O89.3   **Toxic reaction to local anaesthesia during the puerperium**

O89.5   **Other complications of spinal and epidural anaesthesia during the puerperium**

O89.6   **Failed or difficult intubation during the puerperium**

O89.8   **Other complications of anaesthesia during the puerperium**

O89.9   **Complication of anaesthesia during the puerperium, unspecified**

Category: Other obstetric surgery and procedures

Group 7: Non-obstetric complications

O10 **Pre-existing hypertension complicating pregnancy, childbirth and the puerperium**

Incl.: the listed conditions with pre-existing proteinuria

Excl.: that with increased or superimposed proteinuria (O11)

O10.0 **Pre-existing essential hypertension complicating pregnancy, childbirth and the puerperium**

Any condition in I10 specified as a reason for obstetric care during pregnancy, childbirth or the puerperium

O10.1 **Pre-existing hypertensive heart disease complicating pregnancy, childbirth and the puerperium**

Any condition in I11.- specified as a reason for obstetric care during pregnancy, childbirth or the puerperium

O10.2 **Pre-existing hypertensive renal disease complicating pregnancy, childbirth and the puerperium**

Any condition in I12.- specified as a reason for obstetric care during pregnancy, childbirth or the puerperium

O10.3 **Pre-existing hypertensive heart and renal disease complicating pregnancy, childbirth and the puerperium**

Any condition in I13.- specified as a reason for obstetric care during pregnancy, childbirth or the puerperium

O10.4 **Pre-existing secondary hypertension complicating pregnancy, childbirth and the puerperium**

Any condition in I15.- specified as a reason for obstetric care during pregnancy, childbirth or the puerperium

O10.9 **Unspecified pre-existing hypertension complicating pregnancy, childbirth and the Puerperium**

O24 **Diabetes mellitus in pregnancy**

Incl.: in childbirth and the puerperium

O24.0 Pre-existing diabetes mellitus, insulin-dependent

O24.1 Pre-existing diabetes mellitus, non-insulin-dependent

O24.2 Pre-existing malnutrition-related diabetes mellitus

O24.3 Pre-existing diabetes mellitus, unspecified

O24.9 Diabetes mellitus in pregnancy, unspecified

O98 **Maternal infectious and parasitic diseases classifiable elsewhere but complicating pregnancy, childbirth and the puerperium**

Incl.: the listed conditions when complicating the pregnant state, when aggravated by the pregnancy, or as a reason for obstetric care

Use additional code (Chapter I), if desired, to identify specific condition.

Excl.: asymptomatic human immunodeficiency virus [HIV] infection status (Z21)

laboratory evidence of human immunodeficiency virus [HIV] (R75)

obstetrical tetanus (A34)

puerperal:

• infection (O86.-)

• sepsis (O85)

when the reason for maternal care is that the disease is known or suspected to have affected the fetus (O35-O36)

O98.0 Tuberculosis complicating pregnancy, childbirth and the puerperium

Conditions in A15-A19

O98.1 Syphilis complicating pregnancy, childbirth and the puerperium

Conditions in A50-A53

O98.2 Gonorrhoea complicating pregnancy, childbirth and the puerperium

Conditions in A54.-

O98.3 Other infections with a predominantly sexual mode of transmission complicating pregnancy, childbirth and the puerperium

Conditions in A55-A64

O98.4 Viral hepatitis complicating pregnancy, childbirth and the puerperium

Conditions in B15-B19

O98.5 Other viral diseases complicating pregnancy, childbirth and the puerperium

Conditions in A80-B09, B25-B34

O98.6 Protozoal diseases complicating pregnancy, childbirth and the puerperium

Conditions in B50-B64

O98.7 Human immunodeficiency [HIV] disease complicating pregnancy, childbirth and the puerperium

Conditions in (B20-B24)

O98.8 Other maternal infectious and parasitic diseases complicating pregnancy, childbirth and the puerperium

O98.9 Unspecified maternal infectious or parasitic disease complicating pregnancy, childbirth and the puerperium

In all cases O98 must be included in the coding (O98 – Maternal infectious and parasitic diseases classifiable elsewhere but complicating pregnancy, childbirth and the puerperium).

O99 Other maternal diseases classifiable elsewhere but complicating pregnancy, childbirth and the puerperium

Note: This category includes conditions which complicate the pregnant state, are aggravated by the pregnancy or are a main reason for obstetric care and for which the Alphabetical Index does not indicate a specific rubric in .

Use additional code, if desired, to identify specific condition. See below for common codes.

Excl.: infectious and parasitic diseases (O98.-)

injury, poisoning and certain other consequences of external causes (S00-T98)

when the reason for maternal care is that the condition is known or suspected to have affected the fetus (O35-O36)

O99.0 Anaemia complicating pregnancy, childbirth and the puerperium

Conditions in D50-D64

O99.1 Other diseases of the blood and blood-forming organs and certain disorders involving the immune mechanism complicating pregnancy, childbirth and the puerperium

Conditions in D65-D89

Excl.: haemorrhage with coagulation defects (O46.0 , O67.0 , O72.3)

O99.2 Endocrine, nutritional and metabolic diseases complicating pregnancy, childbirth and the puerperium

Conditions in E00-E90

Excl.: diabetes mellitus (O24.-)

malnutrition (O25)

postpartum thyroiditis (O90.5)

O99.3 Mental disorders and diseases of the nervous system complicating pregnancy, childbirth and the puerperium

Conditions in F00-F99 and G00-G99

Excl.: postnatal depression (F53.0)

pregnancy-related peripheral neuritis (O26.8)

puerperal psychosis (F53.1)

O99.4 Diseases of the circulatory system complicating pregnancy, childbirth and the puerperium

Conditions in I00-I99

Excl.: cardiomyopathy in the puerperium (O90.3)

hypertensive disorders (O10-O16)

obstetric embolism (O88.-)

venous complications and cerebrovenous sinus thrombosis in:

• labour, childbirth and the puerperium (O87.-)

• pregnancy (O22.-)

O99.5 Diseases of the respiratory system complicating pregnancy, childbirth and the puerperium

Conditions in J00-J99

O99.6 **Diseases of the digestive system complicating pregnancy, childbirth and the puerperium**

Conditions in K00-K93

Excl.: liver disorders in pregnancy, childbirth and the puerperium (O26.6)

O99.7 **Diseases of the skin and subcutaneous tissue complicating pregnancy, childbirth and the puerperium**

Conditions in L00-L99

Excl.: herpes gestationis (O26.4)

O99.8 **Other specified diseases and conditions complicating pregnancy, childbirth and the puerperium**

Combination of conditions classifiable to O99.0-O99.7

Conditions in C00-D48, H00-H95, M00-M99, N00-N99, and Q00-Q99 not elsewhere classified

Excl.: genitourinary infections in pregnancy (O23.-)

infection of genitourinary tract following delivery (O86.0-O86.3)

maternal care for known or suspected abnormality of maternal pelvic organs (O34.-)

postpartum acute renal failure (O90.4)

postpartum nephritis (O90.8)

Codes for common indirect causes of maternal death

Indirect deaths may also be coded using codes in other chapters of the ICD. However, for correct coding it is essential that reporting on the death certificate mentions clearly in Part 1 that there was mutual aggravation between the condition and the pregnancy. The list below includes common indirect causes of maternal death. Rare diseases are covered by the "Other – specify" label. There is an ICD code for most medical and surgical conditions and coders will follow rules as described in Volume 2 of ICD-10. Countries may wish to expand this list to include disease common to the country or region (e.g. melanoma in Australia, cancer of the cervix in South Africa). :

Diseases of the nervous system		(G00- G99)
	Epilepsy	G40
Diseases of the circulatory system		(I00–I99)
	Rheumatic heart disease	I09
	Bacterial endocarditis	I33
	Artificial valve complications	T82
	Congenital heart disease	(Q20–28)
	Acute myocardial infarction	I21
	Cardiomyopathy	I42
	Cerebrovascular accident[a]	I61
	Stroke	I64
	Other – specify	
	Undiagnosed	I51
Diseases of the respiratory system		(J00–J99)
	Asthma	J85
	Other – specify	
Diseases of the digestive system		(K00–K93)
	Appendicitis	K35
	Pancreatitis	K85
	Liver disease[a] – specify	
	Intestine – specify	
	Other – specify	
Diseases of the musculoskeletal system and connective tissue		(M00–M99)
	Systemic lupus erythematosus	M32
	Kyphoscoliosis	M40
	Other – specify	
Diseases of the genitourinary system		(N00–N99)
	Renal – specify	
	Genital – specify	

[a] Excluding liver disease and cerebral haemorrhage due to pre-eclampsia.
[a] Non-pregnancy-related infection is so named to differentiate this category from infections specific to pregnancy, e.g. puerperal sepsis, septic miscarriage.

Group 8: Unknown/undetermined

O95 **Obstetric death of unspecified cause**
Incl.: Maternal death from unspecified cause occurring during pregnancy, labour and delivery, or the puerperium

Group 9: Coincidental causes

These deaths occur in pregnancy, childbirth, or the puerperium but are not by definition are considered maternal deaths

Coincidental category	Disease entity	ICD-10
	Motor vehicle accident	Y85
		V01–V99
	External causes of accidental injury	W00–Y04
		Y06–Y09
		Y98
	Assault	X85–Y04+
		Y06–Y09
		Y87
	Rape	Y05
	Event of undetermined intent	Y10–Y34
	Other accidents	Y86
	Herbal medication	T65
	Other – specify	

Annex B2: Tabular List of Chapter 15 codes that describe conditions which are unlikely to cause death but may have contributed to the death (contributory condition)

Codes from Annex B1 are preferred. Codes from this block should not be selected as the underlying cause of death if any condition in Annex B1 is present.

When the cause is unspecified, code to "Unknown, Group 8" according to MRG 1244: It is often difficult to identify a maternal death, particularly in cases of indirect obstetric causes. If there is any doubt that the cause of death is obstetrical, for example if the conditions entered in Part 1 are not obstetrical but there is a mention of pregnancy or delivery in Part 2, additional information should be sought from the certifier. This is particularly important in countries where maternal mortality rate is high. If no additional information can be found, deaths with a mention of pregnancy and delivery in Part 1 should be considered obstetrical, but not deaths where pregnancy or delivery is mentioned in Part 2 only.

Further modifications as published in the 11th revision of the ICD may result in changes.

O08 Complications following abortion and ectopic and molar pregnancy

Note: This code is provided primarily for morbidity coding. For use of this category reference should be made to the morbidity coding rules and guidelines in Volume 2.

O08.0 Genital tract and pelvic infection following abortion and ectopic and molar pregnancy

Endometritis

Oophoritis

Parametritis

Pelvic peritonitis

Salpingitis

Salpingo-oophoritis

Sepsis

following conditions classifiable to O00-O07

Use additional code (R57.2), if desired, to identify septic shock.

Excl.: septic or septicopyaemic embolism (O08.2)

urinary tract infection (O08.8)

O08.1 Delayed or excessive haemorrhage following abortion and ectopic and molar pregnancy

Afibrinogenaemia

Defibrination syndrome

Intravascular coagulation

following conditions classifiable to O00-O07

O08.2 **Embolism following abortion and ectopic and molar pregnancy**

Embolism:

- NOS
- air
- amniotic fluid
- blood-clot
- pulmonary
- pyaemic
- septic or septicopyaemic
- soap

following conditions classifiable to O00-O07

O08.3 **Shock following abortion and ectopic and molar pregnancy**

Circulatory collapse

Shock (postoperative) following conditions classifiable to O00-O07

Excl.: septic shock (R57.2)

O08.4 **Renal failure following abortion and ectopic and molar pregnancy**

Oliguria

Renal:

- failure (acute)
- shutdown
- tubular necrosis

Uraemia

following conditions classifiable to O00-O07

O08.5 **Metabolic disorders following abortion and ectopic and molar pregnancy**

Electrolyte imbalance following conditions classifiable to O00-O07

O08.6 **Damage to pelvic organs and tissues following abortion and ectopic and molar pregnancy**

Laceration, perforation, tear or chemical damage of:

- bladder
- bowel
- broad ligament
- cervix
- periurethral tissue
- uterus

following conditions classifiable to O00-O07

O08.7 **Other venous complications following abortion and ectopic and molar pregnancy**

O08.8 **Other complications following abortion and ectopic and molar pregnancy**

Cardiac arrest

Urinary tract infection following conditions classifiable to O00-O07

O08.9	**Complication following abortion and ectopic and molar pregnancy, unspecified**	

Unspecified complication following conditions classifiable to O00-O07

O21.0	**Mild hyperemesis gravidarum**	

Hyperemesis gravidarum, mild or unspecified, starting before the end of the 22nd week of gestation

O21.8 **Other vomiting complicating pregnancy**

Vomiting due to diseases classified elsewhere, complicating pregnancy

Use additional code, if desired, to identify cause.

O21.9 **Vomiting of pregnancy, unspecified**

O22.0 **Varicose veins of lower extremity in pregnancy**

Varicose veins NOS in pregnancy

O22.1 **Genital varices in pregnancy**

Perineal

Vaginal

Vulval

varices in pregnancy

O22.2 **Superficial thrombophlebitis in pregnancy**

Thrombophlebitis of legs in pregnancy

O22.4 **Haemorrhoids in pregnancy**

O25 Malnutrition in pregnancy

Incl.: Malnutrition in childbirth and the puerperium

O26 Maternal care for other conditions predominantly related to pregnancy

O26.0 **Excessive weight gain in pregnancy**

Excl.: gestational oedema (O12.0 , O12.2)

O26.1 **Low weight gain in pregnancy**

O26.2 **Pregnancy care of habitual aborter**

Excl.: habitual aborter:

• with current abortion (O03-O06)

• without current pregnancy (N96)

O26.3 **Retained intrauterine contraceptive device in pregnancy**

O26.4 **Herpes gestationis**

O26.5 **Maternal hypotension syndrome**

Supine hypotensive syndrome

O26.7 Subluxation of symphysis (pubis) in pregnancy, childbirth and the puerperium

Excl.: traumatic separation of symphysis (pubis) during childbirth (O71.6)

O26.8 Other specified pregnancy-related conditions

Exhaustion and fatigue

Peripheral neuritis

Renal disease pregnancy-related

O26.9 Pregnancy-related condition, unspecified

O28 Abnormal findings on antenatal screening of mother

Excl.: diagnostic findings classified elsewhere - see Alphabetical Index maternal care related to the fetus and amniotic cavity and possible delivery problems (O30-O48)

O28.0 Abnormal haematological finding on antenatal screening of mother

O28.1 Abnormal biochemical finding on antenatal screening of mother

O28.2 Abnormal cytological finding on antenatal screening of mother

O28.3 Abnormal ultrasonic finding on antenatal screening of mother

O28.4 Abnormal radiological finding on antenatal screening of mother

O28.5 Abnormal chromosomal and genetic finding on antenatal screening of mother

O28.8 Other abnormal findings on antenatal screening of mother

O28.9 Abnormal finding on antenatal screening of mother, unspecified

O29

O29.4 Spinal and epidural anaesthesia-induced headache during pregnancy

Maternal care related to the fetus and amniotic cavity and possible delivery problems

O30 Multiple gestation

Excl.: complications specific to multiple gestation (O31.-)

O30.0 Twin pregnancy

O30.1 Triplet pregnancy

O30.2 Quadruplet pregnancy

O30.8 Other multiple gestation

O30.9 Multiple gestation, unspecified

Multiple pregnancy NOS

O31 Complications specific to multiple gestation

Excl.: conjoined twins causing disproportion (O33.7) delayed delivery of second twin, triplet, etc. (O63.2)

malpresentation of one fetus or more (O32.5)

with obstructed labour (O64-O66)

O31.0 Papyraceous fetus

Fetus compressus

O31.1 Continuing pregnancy after abortion of one fetus or more

O31.2 Continuing pregnancy after intrauterine death of one fetus or more

O31.8 Other complications specific to multiple gestation

O32 Maternal care for known or suspected malpresentation of fetus

Incl.: the listed conditions as a reason for observation, hospitalization or other obstetric care of the mother, or for caesarean section before onset of labour

Excl.: the listed conditions with obstructed labour (O64.-)

O32.0 Maternal care for unstable lie

O32.1 Maternal care for breech presentation

O32.2 Maternal care for transverse and oblique lie

Presentation:

• oblique

• transverse

O32.3 Maternal care for face, brow and chin presentation

O32.4 Maternal care for high head at term

Failure of head to enter pelvic brim

O32.5 Maternal care for multiple gestation with malpresentation of one fetus or more

O32.6 Maternal care for compound presentation

O32.8 Maternal care for other malpresentation of fetus

O32.9 Maternal care for malpresentation of fetus, unspecified

O33 Maternal care for known or suspected disproportion

Incl.: the listed conditions as a reason for observation, hospitalization or other obstetric care of the mother, or for caesarean section before onset of labour

Excl.: the listed conditions with obstructed labour (O65-O66)

O33.0 Maternal care for disproportion due to deformity of maternal pelvic bones

Pelvic deformity causing disproportion NOS

O33.1 Maternal care for disproportion due to generally contracted pelvis

Contracted pelvis NOS causing disproportion

O33.2 Maternal care for disproportion due to inlet contraction of pelvis

Inlet contraction (pelvis) causing disproportion

O33.3 Maternal care for disproportion due to outlet contraction of pelvis

Mid-cavity contraction (pelvis)

Outlet contraction (pelvis) causing disproportion

O33.4 Maternal care for disproportion of mixed maternal and fetal origin

O33.5 Maternal care for disproportion due to unusually large fetus

Disproportion of fetal origin with normally formed fetus

Fetal disproportion NOS

O33.6   **Maternal care for disproportion due to hydrocephalic fetus**

O33.7   **Maternal care for disproportion due to other fetal deformities**

Conjoined twins

Fetal:

- ascites
- hydrops
- meningomyelocele
- sacral teratoma
- tumour

causing disproportion

O33.8   **Maternal care for disproportion of other origin**

O33.9   **Maternal care for disproportion, unspecified**

Cephalopelvic disproportion NOS

Fetopelvic disproportion NOS

O34   Maternal care for known or suspected abnormality of pelvic organs

Incl.: the listed conditions as a reason for observation, hospitalization or other obstetric care of the mother, or for caesarean section before onset of labour

Excl.: the listed conditions with obstructed labour (O65.5)

O34.0   **Maternal care for congenital malformation of uterus**

Maternal care for:

- double uterus
- uterus bicornis

O34.1   **Maternal care for tumour of corpus uteri**

Maternal care for:

- polyp of corpus uteri
- uterine fibroid

Excl.: maternal care for tumour of cervix (O34.4)

O34.2   **Maternal care due to uterine scar from previous surgery**

Maternal care for scar from previous caesarean section

Excl.: vaginal delivery following previous caesarean section NOS (O75.7)

O34.3   **Maternal care for cervical incompetence**

Maternal care for:

- cerclage
- Shirodkar suture

with or without mention of cervical incompetence

O34.4   **Maternal care for other abnormalities of cervix**

Maternal care for:

- polyp of cervix
- previous surgery to cervix
- stricture or stenosis of cervix
- tumour of cervix

O34.5 **Maternal care for other abnormalities of gravid uterus**

Maternal care for:

- incarceration
- prolapse
- retroversion

of gravid uterus

O34.6 **Maternal care for abnormality of vagina**

Maternal care for:

- previous surgery to vagina
- septate vagina
- stenosis of vagina (acquired)(congenital)
- stricture of vagina
- tumour of vagina

Excl.: maternal care for vaginal varices in pregnancy (O22.1)

O34.7 **Maternal care for abnormality of vulva and perineum**

Maternal care for:

- fibrosis of perineum
- previous surgery to perineum or vulva
- rigid perineum
- tumour of vulva

Excl.: maternal care for perineal and vulval varices in pregnancy (O22.1)

O34.8 **Maternal care for other abnormalities of pelvic organs**

Maternal care for:

- cystocele
- pelvic floor repair (previous)
- pendulous abdomen
- rectocele
- rigid pelvic floor

O34.9 **Maternal care for abnormality of pelvic organ, unspecified**

O35 Maternal care for known or suspected fetal abnormality and damage

Incl.: the listed conditions in the fetus as a reason for observation, hospitalization or other obstetric care of the mother, or for termination of pregnancy

Excl.: maternal care for known or suspected disproportion (O33.-)

O35.0 **Maternal care for (suspected) central nervous system malformation in fetus**

Maternal care for (suspected) fetal:

- anencephaly
- spina bifida

Excl.: chromosomal abnormality in fetus (O35.1)

O35.1 **Maternal care for (suspected) chromosomal abnormality in fetus**

O35.2 **Maternal care for (suspected) hereditary disease in fetus**

Excl.: chromosomal abnormality in fetus (O35.1)

O35.3 **Maternal care for (suspected) damage to fetus from viral disease in mother**

Maternal care for (suspected) damage to fetus from maternal:
- cytomegalovirus infection
- rubella

O35.4 **Maternal care for (suspected) damage to fetus from alcohol**

O35.5 **Maternal care for (suspected) damage to fetus by drugs**

Maternal care for (suspected) damage to fetus from drug addiction

Excl.: fetal distress in labour and delivery due to drug administration (O68.-)

O35.6 **Maternal care for (suspected) damage to fetus by radiation**

O35.7 **Maternal care for (suspected) damage to fetus by other medical procedures**

Maternal care for (suspected) damage to fetus by:
- amniocentesis
- biopsy procedures
- haematological investigation
- intrauterine contraceptive device
- intrauterine surgery

O35.8 **Maternal care for other (suspected) fetal abnormality and damage**

Maternal care for (suspected) damage to fetus from maternal:
- listeriosis
- toxoplasmosis

O35.9 **Maternal care for (suspected) fetal abnormality and damage, unspecified**

O36 Maternal care for other known or suspected fetal problems

Incl.: the listed conditions in the fetus as a reason for observation, hospitalization or other obstetric care of the mother, or for termination of pregnancy

Excl.: labour and delivery complicated by fetal stress [distress] (O68.-)

placental transfusion syndromes (O43.0)

O36.0 **Maternal care for rhesus isoimmunization**

Anti-D [Rh] antibodies

Rh incompatibility (with hydrops fetalis)

O36.1 **Maternal care for other isoimmunization**

ABO isoimmunization

Isoimmunization NOS (with hydrops fetalis)

O36.2 **Maternal care for hydrops fetalis**

Hydrops fetalis:
- NOS
- not associated with isoimmunization

O36.3 **Maternal care for signs of fetal hypoxia**

O36.4 **Maternal care for intrauterine death**

Excl.: missed abortion (O02.1)

O36.5 **Maternal care for poor fetal growth**

Maternal care for known or suspected:

- light-for-dates
- placental insufficiency
- small-for-dates

O36.6 **Maternal care for excessive fetal growth**

Maternal care for known or suspected large-for-dates

O36.7 **Maternal care for viable fetus in abdominal pregnancy**

O36.8 **Maternal care for other specified fetal problems**

O36.9 **Maternal care for fetal problem, unspecified**

O40 Polyhydramnios

Incl.: Hydramnios

O41 Other disorders of amniotic fluid and membranes

Excl.: premature rupture of membranes (O42.-)

O41.0 **Oligohydramnios**

Oligohydramnios without mention of rupture of membranes

O41.8 **Other specified disorders of amniotic fluid and membranes**

O41.9 **Disorder of amniotic fluid and membranes, unspecified**

O42 Premature rupture of membranes

O42.0 **Premature rupture of membranes, onset of labour within 24 hours**

O42.1 **Premature rupture of membranes, onset of labour after 24 hours**

Excl.: with labour delayed by therapy (O42.2)

O42.2 **Premature rupture of membranes, labour delayed by therapy**

O42.9 **Premature rupture of membranes, unspecified**

O43 Placental disorders

Excl.: maternal care for poor fetal growth due to placental insufficiency (O36.5)

placenta praevia (O44.-)

premature separation of placenta [abruptio placentae] (O45.-)

O43.0 **Placental transfusion syndromes**

Transfusion:

- fetomaternal
- maternofetal
- twin-to-twin

O43.1 **Malformation of placenta**

Abnormal placenta NOS

Circumvallate placenta

O43.8 Other placental disorders

Placental:

- dysfunction
- infarction

O43.9 Placental disorder, unspecified

O47 False labour

O47.0 False labour before 37 completed weeks of gestation

O47.1 False labour at or after 37 completed weeks of gestation

O47.9 False labour, unspecified

O48 Prolonged pregnancy

Incl.: Post-dates

Post-term

O60 Preterm labour and delivery

Incl.: Onset (spontaneous) of labour before 37 completed weeks of gestation

O60.0 Preterm labour without delivery

Preterm labour:

- induced
- spontaneous

O60.1 Preterm spontaneous labour with preterm delivery

Preterm labour with delivery NOS

Preterm spontaneous labour with preterm delivery by caesarean section

O60.2 Preterm labour with term delivery

Preterm spontaneous labour with term delivery by caesarean section

O60.3 Preterm delivery without spontaneous labour

Preterm delivery by:

- caesarean section, without spontaneous labour
- induction

O61 Failed induction of labour

O61.0 Failed medical induction of labour

Failed induction (of labour) by:

- oxytocin
- prostaglandins

O61.1 Failed instrumental induction of labour

Failed induction (of labour):

- mechanical
- surgical

| O61.8 | Other failed induction of labour |
| O61.9 | Failed induction of labour, unspecified |

O62 Abnormalities of forces of labour

O62.0 **Primary inadequate contractions**

Failure of cervical dilatation

Primary hypotonic uterine dysfunction

Uterine inertia during latent phase of labour

O62.1 **Secondary uterine inertia**

Arrested active phase of labour

Secondary hypotonic uterine dysfunction

O62.2 **Other uterine inertia**

Atony of uterus

Desultory labour

Hypotonic uterine dysfunction NOS

Irregular labour

Poor contractions

Uterine inertia NOS

O62.3 **Precipitate labour**

O62.4 **Hypertonic, incoordinate, and prolonged uterine contractions**

Contraction ring dystocia

Dyscoordinate labour

Hour-glass contraction of uterus

Hypertonic uterine dysfunction

Incoordinate uterine action

Tetanic contractions

Uterine dystocia NOS

Excl.: dystocia (fetal)(maternal) NOS (O66.9)

O62.8 **Other abnormalities of forces of labour**

O62.9 **Abnormality of forces of labour, unspecified**

O63 Long labour

O63.0 **Prolonged first stage (of labour)**

O63.1 **Prolonged second stage (of labour)**

O63.2 **Delayed delivery of second twin, triplet, etc.**

O63.9 **Long labour, unspecified**

Prolonged labour NOS

NOTE: codes O64 to O66, are contributory conditions and alone do not provide insufficient detail on the cause of death. In cases where O64 - O66 are indicated as cause of death, these will be tabulated under "other obstetric causes" and considered a direct maternal death.

O64 Obstructed labour due to malposition and malpresentation of fetus

O64.0 Obstructed labour due to incomplete rotation of fetal head

Deep transverse arrest

Obstructed labour due to persistent (position):

- occipitoiliac
- occipitoposterior
- occipitosacral
- occipitotransverse

O64.1 Obstructed labour due to breech presentation

O64.2 Obstructed labour due to face presentation

Obstructed labour due to chin presentation

O64.3 Obstructed labour due to brow presentation

O64.4 Obstructed labour due to shoulder presentation

Excl.: impacted shoulders (O66.0)

shoulder dystocia (O66.0)

O64.5 Obstructed labour due to compound presentation

O64.8 Obstructed labour due to other malposition and malpresentation

O64.9 Obstructed labour due to malposition and malpresentation, unspecified

O65 Obstructed labour due to maternal pelvic abnormality

O65.0 Obstructed labour due to deformed pelvis

O65.1 Obstructed labour due to generally contracted pelvis

O65.2 Obstructed labour due to pelvic inlet contraction

O65.3 Obstructed labour due to pelvic outlet and mid-cavity contraction

O65.4 Obstructed labour due to fetopelvic disproportion, unspecified

Excl.: dystocia due to abnormality of fetus (O66.2-O66.3)

O65.5 Obstructed labour due to abnormality of maternal pelvic organs

Obstructed labour due to conditions listed in O34.-

O65.8 Obstructed labour due to other maternal pelvic abnormalities

O65.9 Obstructed labour due to maternal pelvic abnormality, unspecified

O66 Other obstructed labour

O66.0 Obstructed labour due to shoulder dystocia

Impacted shoulders

O66.1 Obstructed labour due to locked twins

O66.2 Obstructed labour due to unusually large fetus

O66.3 Obstructed labour due to other abnormalities of fetus

Dystocia due to:

- conjoined twins
- fetal

- ascites

- hydrops

- meningomyelocele

- sacral teratoma

- tumour

- hydrocephalic fetus

O66.4 **Failed trial of labour, unspecified**

Failed trial of labour with subsequent delivery by caesarean section

O66.5 **Failed application of vacuum extractor and forceps, unspecified**

Failed application of ventouse or forceps, with subsequent delivery by forceps or caesarean section

Respectively

O66.8 **Other specified obstructed labour**

O66.9 **Obstructed labour, unspecified**

Dystocia:

- NOS

- fetal NOS

- maternal NOS

O68 Labour and delivery complicated by fetal stress [distress]

Incl.: fetal distress in labour or delivery due to drug administration

O68.0 **Labour and delivery complicated by fetal heart rate anomaly**

Fetal:

- bradycardia

- heart rate irregularity

- tachycardia

Excl.: with meconium in amniotic fluid (O68.2)

O68.1 **Labour and delivery complicated by meconium in amniotic fluid**

Excl.: with fetal heart rate anomaly (O68.2)

O68.2 **Labour and delivery complicated by fetal heart rate anomaly with meconium in amniotic fluid**

O68.3 **Labour and delivery complicated by biochemical evidence of fetal stress**

Abnormal fetal:

- acidaemia

- acid-base balance

O68.8 **Labour and delivery complicated by other evidence of fetal stress**

Evidence of fetal distress:

- electrocardiographic

- ultrasonic

O68.9 **Labour and delivery complicated by fetal stress, unspecified**

O69	Labour and delivery complicated by umbilical cord complications
O69.0	Labour and delivery complicated by prolapse of cord
O69.1	Labour and delivery complicated by cord around neck, with compression
O69.2	Labour and delivery complicated by other cord entanglement, with compression

    Compression of cord NOS

    Entanglement of cords of twins in monoamniotic sac

    Knot in cord

O69.3	Labour and delivery complicated by short cord
O69.4	Labour and delivery complicated by vasa praevia

    Haemorrhage from vasa praevia

O69.5	Labour and delivery complicated by vascular lesion of cord

    Cord:

    • bruising

    • haematoma

    Thrombosis of umbilical vessels

O69.8	Labour and delivery complicated by other cord complications

    Cord around neck without compression

O69.9	Labour and delivery complicated by cord complication, unspecified

O70	Perineal laceration during delivery

Incl.: episiotomy extended by laceration

Excl.: obstetric high vaginal laceration alone (O71.4)

O70.0	First degree perineal laceration during delivery

    Perineal laceration, rupture or tear (involving):

    • fourchette

    • labia

    • skin

    • slight

    • vagina

    • vulva

    during delivery

O70.1	Second degree perineal laceration during delivery

    Perineal laceration, rupture or tear as in O70.0, also involving:

    • pelvic floor

    • perineal muscles

    • vaginal muscles

    during delivery

Excl.: that involving anal sphincter (O70.2)

O70.2 **Third degree perineal laceration during delivery**

Perineal laceration, rupture or tear as in O70.1, also involving:

- anal sphincter
- rectovaginal septum
- sphincter NOS

during delivery

Excl.: that involving anal or rectal mucosa (O70.3)

O70.3 **Fourth degree perineal laceration during delivery**

Perineal laceration, rupture or tear as in O70.2, also involving:

- anal mucosa
- rectal mucosa

during delivery

O70.9 **Perineal laceration during delivery, unspecified**

O74.5 **Spinal and epidural anaesthesia-induced headache during labour and delivery**

O75 Other complications of labour and delivery, not elsewhere classified

Excl.: puerperal:
- infection (O86.-)
- sepsis (O85)

O75.0 **Maternal distress during labour and delivery**

O75.1 **Shock during or following labour and delivery**

Obstetric shock

O75.2 **Pyrexia during labour, not elsewhere classified**

O75.5 **Delayed delivery after artificial rupture of membranes**

O75.6 **Delayed delivery after spontaneous or unspecified rupture of membranes**

Excl.: spontaneous premature rupture of membranes (O42.-)

O75.7 **Vaginal delivery following previous caesarean section**

Delivery (O80–O84)

Note: For use of these categories reference should be made to the mortality and morbidity coding rules and guidelines in Volume 2.

O80 Single spontaneous delivery

Incl.: cases with minimal or no assistance, with or without episiotomy delivery in a completely normal case

- O80.0 Spontaneous vertex delivery
- O80.1 Spontaneous breech delivery
- O80.8 Other single spontaneous delivery
- O80.9 Single spontaneous delivery, unspecified

  Spontaneous delivery NOS

O81 Single delivery by forceps and vacuum extractor

Excl.: failed application of vacuum extractor or forceps (O66.5)

- O81.0 Low forceps delivery
- O81.1 Mid-cavity forceps delivery
- O81.2 Mid-cavity forceps with rotation
- O81.3 Other and unspecified forceps delivery
- O81.4 Vacuum extractor delivery

  Ventouse delivery

- O81.5 Delivery by combination of forceps and vacuum extractor

  Forceps and ventouse delivery

O82 Single delivery by caesarean section

- O82.0 Delivery by elective caesarean section

  Repeat caesarean section NOS

- O82.1 Delivery by emergency caesarean section
- O82.2 Delivery by caesarean hysterectomy
- O82.8 Other single delivery by caesarean section
- O82.9 Delivery by caesarean section, unspecified

O83 Other assisted single delivery

- O83.0 Breech extraction
- O83.1 Other assisted breech delivery

  Breech delivery NOS

- O83.2 Other manipulation-assisted delivery

  Version with extraction

- O83.3 Delivery of viable fetus in abdominal pregnancy

O83.4	**Destructive operation for delivery**

Cleidotomy

Craniotomy

Embryotomy

to facilitate delivery

O83.8	**Other specified assisted single delivery**
O83.9	**Assisted single delivery, unspecified**

Assisted delivery NOS

O84 Multiple delivery

Use additional code (O80-O83), if desired, to indicate the method of delivery of each fetus or infant.

O84.0	**Multiple delivery, all spontaneous**
O84.1	**Multiple delivery, all by forceps and vacuum extractor**
O84.2	**Multiple delivery, all by caesarean section**
O84.8	**Other multiple delivery**

Multiple delivery by combination of methods

O84.9	**Multiple delivery, unspecified**

O87.0	**Superficial thrombophlebitis in the puerperium**
O87.2	**Haemorrhoids in the puerperium**
O87.8	**Other venous complications in the puerperium**

Genital varices in the puerperium

O89.4	**Spinal and epidural anaesthesia-induced headache during the puerperium**

O92 Other disorders of breast and lactation associated with childbirth

Incl.: the listed conditions during pregnancy, the puerperium or lactation

O92.0	**Retracted nipple associated with childbirth**
O92.1	**Cracked nipple associated with childbirth**

Fissure of nipple, gestational or puerperal

O92.2	**Other and unspecified disorders of breast associated with childbirth**
O92.3	**Agalactia**

Primary agalactia

O92.4	**Hypogalactia**
O92.5	**Suppressed lactation**

Agalactia:

- elective
- secondary
- therapeutic

O92.6	**Galactorrhoea**

Excl.: galactorrhoea not associated with childbirth (N64.3)

O92.7	**Other and unspecified disorders of lactation**

Puerperal galactocele

Other obstetric conditions, not elsewhere classified (O94-O99)

Note: For use of categories O95-O97 reference should be made to the mortality coding rules and guidelines in Volume 2.

O94 Sequelae of complication of pregnancy, childbirth and the puerperium

Note: This category is to be used for morbidity coding only to indicate conditions in categories O00-O75 and O85-O92 as the cause of sequelae, which are themselves classified elsewhere.

The 'sequelae' include conditions specified as such or as late effects, or those present one year or more after the onset of the causal condition.

Not to be used for chronic complications of pregnancy, childbirth and the puerperium. Code these to O00-O75 and O85-O92 .

Excl.: that resulting in death (O96.- , O97.-)

Annex B3: Tabular List of Other codes of interest

O96 Death from any obstetric cause occurring more than 42 days but less than one year after delivery

Use additional code, if desired, to identify obstetric cause (direct or indirect) of death.

- O96.0 Death from direct obstetric cause
- O96.1 Death from indirect obstetric cause
- O96.9 Death from unspecified obstetric cause

O97 Death from sequelae of obstetric causes

Incl.: Death from any obstetric cause (direct or indirect) occurring one year or more after delivery. Use additional code, if desired, to identify the obstetric cause (direct or indirect) (O43-O45)

- O97.0 Death from sequelae of direct obstetric cause
- O97.1 Death from sequelae of indirect obstetric cause
- O97.9 Death from sequelae of obstetric cause, unspecified

Annex C: Suggestions of tools and examples to facilitate the implementation of the guide and its groupings

Once the underlying causes/disease entities have been defined and identified, the contributory conditions need to be listed.

Countries may include a checklist of relevant contributory conditions be included on the maternal death data sheet as illustrated below. The conditions could be marked in the appropriate box and then all the deaths with the condition could be selected. One would thereby have all the disease entities that resulted from that contributory condition. Countries could select which conditions are relevant to them and include them so the list does not become exhaustive.

Associated/contributory condition checklist

	Yes	No	Unknown	N/A
HIV infection				
Anaemia				
Previous caesarean section				
Prolonged labour				
Female genital mutilation				
Obesity				
Depression				
Domestic violence				
etc.				

Other important information to be collected for any maternal death should be:

- *mode of delivery*: undelivered, normal vaginal, assisted vaginal, caesarean section, abortion/miscarriage
- *pregnancy outcome*: miscarriage/abortion, antenatal death, intrapartum death, neonatal death, alive.